Explaining Culture

Explaining Culture

A Naturalistic Approach

Dan Sperber

BLACKWELL
Publishers

WB

First published 1996
Reprinted 1997

Blackwell Publishers Ltd
108 Cowley Road, Oxford OX4 1JF, UK

Blackwell Publishers Inc
350 Main Street, Malden, Massachusetts 02148, USA

British Library Cataloguing in Publication Data
A CIP catalogue record for this book is available from the British Library

Library of Congress Cataloging in Publication Data
Sperber, Dan
Explaining culture: a naturalistic approach / Dan Sperber.
p.   cm.
Includes bibliographical references and index.
ISBN 0-631-20044-4 (hbk : alk. paper)
ISBN 0-631-20045-2 (pbk : alk. paper)
1. Culture. 2. Cognitive psychology. 3. Social evolution.
4. Cognition and culture. I. Title.
GN357.S66   1996   96-6454
306—dc20   CIP

Typeset in 11 on 13pt Bembo
by Wearset, Boldon, Tyne and Wear
Printed and bound in Great Britain
by Hartnolls Ltd, Bodmin, Cornwall

This book is printed on acid-free paper

9/4/03

Contents

Preface

A spectre haunts the social sciences, the spectre of a natural science of the social. Some wait for the day that the spectre will make itself known, and will at last make the social sciences truly scientific. Others denounce the threat of scientism and reductionism. Some say they speak for the spectre. Others say it is just a hoax. Here is what I think: in lieu of a spectre, there is just a child in limbo. A naturalistic programme in the social sciences is conceivable, but it has yet to be developed. In this book, I present a fragment of such a programme: a naturalistic approach to culture.

The six essays collected in this book are all arguments for, and contributions to, an epidemiology of representations. They were written at different stages in my work over the past ten years, but they were all, in my mind, parts of a single project. After my *On Anthropological Knowledge* (published in French in 1982, and in English in 1985), which was more on the critical side, I wanted to contribute positively to the 'rethinking of anthropology' advocated by Edmund Leach in his famous inaugural Malinowski Memorial Lecture of 1959.

The lectures on which these chapters are based have been presented to a variety of audiences: anthropologists, archaeologists, scholars in the humanities, philosophers, developmental psychologists, and social psychologists. The chapters do not presuppose any specialized competences on the part of readers. Chapter 1 is a new synthesis of two earlier essays, and chapter 5 is entirely new. The four other chapters are based on previously published material. Conceived in relationship to one another and written separately, these chapters have been revised to form a whole. They can be read together or, if one prefers, independently of each other, since each recapitulates the basic ideas that link it to the project as a whole.

The best part of my work, during the past fifteen years, has been done with Deirdre Wilson, and has been devoted to developing relevance theory as a theory of human communication and as a general approach to many issues in cognition. My initial interest in our collaborative project had to do with the role communication plays in culture. One of my goals will be to spell out the implications of relevance theory for the epidemiology of representations.

Over the years, many people have helped me with their advice, criticism, and encouragement, in particular Daniel Andler, Robert Axelrod, Maurice Bloch, Radu Bogdan, Francesco Cara, Philip Carpenter, Jean-Pierre Changeux, Bernard Conein, Leda Cosmides, Helena Cronin, Daniel Dennett, Frank Döring, Jean-Pierre Dupuy, Catherine Elgin, Heidi Feldman, Allan Gibbard, Margaret Gilbert, Vittorio Girotto, Jack Goody, Gilbert Harman, Odile Jacob, Pierre Jacob, Gérard Jorland, Jerry Katz, Helen Lees, Richard Nisbett, Gloria Origgi, David Premack, François Recanati, Jenka Sperber, John Tooby, Jean van Altena, Deirdre Wilson, and three anonymous reviewers. The ideas developed here were first discussed with Scott Atran, Pascal Boyer, and Larry Hirschfeld; their comments have always been particularly helpful, and my work relates to theirs in many obvious ways. Monique Canto-Sperber has had, in all these years, the most welcome influence, not just on my work, but on my life.

I thank you all.

Introduction

The central theme of this book is quite simple. Our individual brains are each inhabited by a large number of ideas that determine our behaviour. Thus my brain is inhabited by (among others) ideas about culture that caused me to write this book. Some of the behaviours of an individual, or some of the traces left by these behaviours in the environment, are observed by others. Here you are, reading this page, which is a trace of my work. Observing a behaviour or its traces gives rise to ideas, such as the ideas that are at this very moment coming to your mind. Sometimes, the ideas caused by a behaviour resemble the ideas that have caused this behaviour. This will be the case if I succeed in making myself understood.

Through a material process like the one just evoked, an idea, born in the brain of one individual, may have, in the brains of other individuals, descendants that resemble it. Ideas can be transmitted, and, by being transmitted from one person to another, they may even propagate. Some ideas – religious beliefs, cooking recipes, or scientific hypotheses, for instance – propagate so effectively that, in different versions, they may end up durably invading whole populations. Culture is made up, first and foremost, of such contagious ideas. It is made up also of all the productions (writings, artworks, tools, etc.) the presence of which in the shared environment of a human group permits the propagation of ideas.

To explain culture, then, is to explain why and how some ideas happen to be contagious. This calls for the development of a true *epidemiology of representations*.

The word 'epidemiology' comes from the Greek *epidemia* which meant 'stay or arrival in a country'. In its most common use, *epidemia* (as well as related words) referred to the stay or arrival of people, but it could also refer to the stay or arrival of things such as rain, diseases, or even customs. Comparing the spread of diseases with that of ideas is an old commonplace, and the word 'contagion' is so frequently used for mental states that the metaphorical character of this usage is barely recognizable anymore. Similarly, the use of 'epidemiology' for a study of the distribution of mental states in a population is a barely metaphorical extension of the term.

Though the word 'epidemiology' is long and rare, the idea it expresses is simple and general. Say you have a population (for instance, a human group) and some interesting property (for instance, being diabetic, having white hair, or believing in witches) that the members of this population may or may not have. An epidemiological approach would consist in describing and explaining the distribution of this property in this population. Epidemiology is not restricted to contagious diseases: diabetes is not contagious, believing in witches is not a disease, and having white hair is neither.

Epidemiology is eclectic in its use of explanatory models. Some models are borrowed from population genetics, some from ecology, some from social psychology, and novel ones may be developed as the need arises. I have chosen the term 'epidemiology' precisely because of this generality and eclecticism. A naturalistic approach to culture requires considering the distribution of a variety of mental and environmental phenomena. For this, different causal models are needed simultaneously.

All epidemiological models, whatever their differences, have in common the fact that they explain population-scale macro-phenomena, such as epidemics, as the cumulative effect of micro-processes that bring about individual events, such as catching a disease. In this, epidemiological models contrast starkly with 'holistic' explanations, in which macro-phenomena are explained in terms of other macro-phenomena – for instance, religion in terms of economic structure (or conversely).

While the idea of cultural contagion is ancient, the first serious attempt at a scientific cultural epidemiology is probably to be found in the work of the French sociologist Gabriel Tarde (see *Les Lois de l'imitation*, 1890). Though he himself hardly used the epidemiological idiom, he insisted that culture – and indeed social life in general – had to be explained as the cumulative effect of countless processes of inter-individual transmission through imitation.

More recently, a number of authors – in particular Donald Campbell (1974), Richard Dawkins (1976, 1982), Cavalli-Sforza and Feldman (1981), Lumsden and Wilson (1981), Boyd and Richerson (1985), and William Durham (1991) – have adapted the Darwinian model of selection to the case of culture. These are epidemiological approaches (so named by Cavalli-Sforza and Feldman, merely described as 'evolutionary' by others). Richard Dawkins has popularized the idea that culture is made up of units he calls 'memes', which, like genes, undergo replication and selection. These Darwinian approaches, which borrow their models from population genetics, grant only a limited role to psychology. Yet the micro-mechanisms that bring about the propagation of ideas are mostly psychological and, more specifically, cognitive mechanisms. Cognitive psychology has undergone unprecedented developments in the past three decades. It has recently benefited from a Darwinian perspective on the psychological evolution of the human species developed independently of Darwinian approaches to culture (see Cosmides and Tooby 1987). I believe that cognitive psychology provides one of the main sources of insight for explaining culture. The approach to culture advocated here is both epidemiological and cognitive, and, as will be seen, is more closely linked to Darwinism on the cognitive side than on the epidemiological side.

I think of the epidemiology of representations as a naturalistic research programme in the social sciences. The social sciences are a large, loose alliance of research programmes with very different goals. They range from socio-linguistics to world trade economics, from legal history to ethno-psychiatry, from the study of Vedic texts to that of voters' choices. Many social science programmes have regional or historical topics. Many are driven by practical concerns. Even a single field such as anthropology (on which I will focus), includes research programmes on topics as diverse as kinship semantics and fishing technology, the postmodern study of post-colonialism, the cultural study of science, nutritional anthropology, and the anthropology of consciousness.

Almost all research programmes in the social sciences insist on the label 'science', if only because science is where research money goes. To insist that they are not *really* sciences, as if 'science' were a well-defined label, the champagne of intellectual products, is often just a devious way of denying them respectability and resources. Though not all research programmes are equally worth encouraging, attempts at belittling the social sciences in general ignore the difficulty of their tasks, the competencies they have accumulated, and the role they play in democratic life. Let social scientists use the label 'science' freely. The interesting question is not whether the social sciences are sciences, but whether they are continuous with the natural sciences (assuming, as I do, that the natural sciences are roughly continuous with one another).

Research programmes in the social sciences tend to exhibit a healthy eclecticism in their methodology, availing themselves of whatever tools may help. In particular, when it pays to use methods borrowed from the natural sciences, they generally do. But, often enough, natural science methodology is cumbersome and unhelpful in the pursuit of social-scientific goals. Psychological imagination, common-sense comprehension, and experience-based expertise are often the most effective tools.

Anyhow, the use of natural science methods may be necessary, but is not sufficient to make a research programme a natural-scientific one (as illustrated by the case of economics: very scientific in its methods, but not at all naturalistic). What matters most is the goal. A prototypical natural-scientific goal is to discover some natural mechanism that explains a wide range of phenomena in a testable manner. Few research programmes in the social sciences have that kind of a goal. Those that do and are reasonably successful at it – as, for instance, in demographic history – are about some very specific aspects of the social domain. I am not aware of any naturalistic programme effectively outlining a causal-mechanistic approach to social phenomena in general.

Why is there no natural science of the social to date? First, because few social scientists have ever cared about developing such a science. Second, and more importantly, because the things the social sciences are about – such as politics, law, religion, money, and art – do not fit in any obvious way into the natural world.

How could one go about trying to fit social things into nature, in other words, 'naturalize' them? Here, cognitive science is relevant in more ways than one. A naturalistic programme is one that establishes fundamental continuities between its domain and that of one or several neighbouring natural sciences. Psychological sciences are the social sciences' closest neighbours, and some of their programmes – roughly those falling under the 'cognitive science' label – are in the process of being more or less successfully naturalized. Naturalizing the social domain would presumably involve establishing some continuity with programmes in cognitive science.

The development of cognitive science has put in a new light the question, 'How do mental things fit into nature?' While it has not yet been answered to everybody's satisfaction, this question is at least much better understood than that of the place of social things in nature. It has been approached in three main ways. The first is to try to *reduce* the mental to the neurological, the natural character of which is uncontroversial. According to reductionism, any description of a mental phenomenon in psychological terms could be translated into neurological terms. The second is to *grant naturalness more liberally*. It could be maintained that every token of a mental phenomenon is a neurological, hence natural, phenomenon, even if its description by means of psychological categories cannot be translated into neurological categories. Thus we have a kind of minimal naturalism, without reductionism. The third way of naturalizing the mental is to *reconceptualize* the whole domain, and eliminate all concepts that do not refer to natural entities. This is generally described as eliminativism.

In the same spirit, three ways of naturalizing the social domain come to mind. Each has its difficulties, however. One might *reduce the social to the natural*. Are social things as we know them reducible to natural things? True reductions are great scientific achievements. In the social sciences, however, 'reductionist' is used as a word of abuse, as if reduction were a genuine option, which had, for some reason, to be forestalled. In fact, no serious reduction of social science concepts or theories to natural science ones has ever been suggested, let alone developed. Reduction, then, is a possibility in principle – salient because of the role it has played in other sciences, and even more in philosophy of science – but not much of a real hope or threat in the present case.

Or one might *grant naturalness more liberally*. It could be argued (adapting ideas of Putnam and Fodor in the philosophy of mind) that every token of a social thing is a token of a natural thing, even if the types, or categories, of sociology are irreducible to those of any other science. But how would this help formulate the kind of generalizations a naturalistic social science should produce? Generalizations are perforce about types, not tokens.

The third possibility is to *reconceptualize the social domain*. It could be that the way we carve the social domain misses its natural joints. If so, then today's social science concepts must be replaced – at least for the purpose of a naturalistic research programme – by a new battery of concepts. The social domain must be carved differently, in such a way that the categories of social things that we recognize become clearly natural categories. But how would you do that? And supposing you knew how to reconceptualize the domain, how would you retain the benefit of the competence expressed by means of the old, now eliminated, conceptual scheme?

If these are the only three conceivable ways to naturalize the social domain, and if they are all beset by such difficulties, why not give up the whole idea? Why shouldn't the social sciences keep to themselves?

Reduction seems to me impossible, and a more liberal granting of the term 'natural', vacuous (in the case of the social sciences, at least; the case of the cognitive sciences is different – see chapter 1). I believe, on the other hand, that the epidemiological approach makes it possible – and even necessary – to reconceptualize the social domain. My proposal is as modest as is possible for such an intrinsically ambitious project. The new conceptual scheme bears, as will be shown, a systematic relation to the standard one, and this allows it to draw extensively on past achievements in the social sciences. The goal of the naturalistic programme turns out to be not a Grand Theory – a physics of the social world, as Auguste Comte imagined – but a complex of interlocking, middle-range models.

Human social life is just one aspect of the life of one animal species among millions, on a little planet somewhere. It is the outcome of the improbable conjunction of countless miscellaneous factors. There is no reason to expect human social life to exhibit the simplicity and systematicity found in physics or chemistry or, to a lesser extent, molecular biology. Many natural sciences – geography, climatology, epidemiology, for instance – have rather messy domains and no Grand Theory. This would be the case also with a natural social science understood as an epidemiology of representations.

Chapter 1, 'How to be a True Materialist in Anthropology', introduces the project of an epidemiology of representations from a philosophical point of view. A naturalistic programme in the social sciences can be pursued, but it requires rethinking the very categories by means of which we approach the domain.

Chapter 2, 'Interpreting and Explaining Cultural Representations', introduces the project from a more social-scientific point of view. It considers the different types of understanding worth aiming at in anthropology. It contrasts, in particular, interpretive and causal explanations. It locates the epidemiological project among other types of causal explanation.

Chapter 3, 'Anthropology and Psychology: Towards an Epidemiology of Representations', expands the general idea of an epidemiology of representations introduced in the two preceding chapters, and illustrates it briefly. It was first delivered as a Malinowski Memorial Lecture at the London School of Economics in 1984, and has often been referred to since. I have therefore made only minor revisions to the original text.

Chapter 4, 'The Epidemiology of Beliefs', develops one of the themes of the preceding chapter, and illustrates how psychology and anthropology may be highly relevant to one another both in answering some of their respective traditional questions and in formulating new common questions. It draws on earlier work of mine on 'apparently irrational beliefs' (see Sperber 1985b: ch. 2) and integrates it within an epidemiological perspective.

While the central idea of an epidemiology of representations is relatively simple to explain, some of the main issues which this approach should help illuminate are fairly complex. The last two chapters deal with such issues, and are even more ambitious than the four preceding ones, but also a bit more difficult.

Chapter 5, 'Selection and Attraction in Cultural Evolution', is about the different ways of modelling cultural evolution. I contrast the selectionist models of cultural evolution defended by Richard Dawkins and others with a more general epidemiological model of 'cultural attraction', in which a greater role is given to psychological mechanisms.

Chapter 6, 'Mental Modularity and Cultural Diversity', takes as its starting-point an idea suggested long ago by Noam Chomsky (whom I echoed on this point in Sperber 1968 and other early writings). The human mind, Chomsky argued, is better viewed not as general all-purpose intelligence, but as a combination of many devices that are in part genetically programmed. These 'modules' (to use the term made popular by Jerry Fodor) are differently specialized, in terms of both the cognitive domains they handle and the type of information processing they perform. There is a tension, however, between this rather nativisit view of cognition and the recognition of human cultural diversity, which suggests, to the contrary, that the mind is indefinitely malleable. One way of resolving the tension would be to deny, or downplay, the modularity of the mind. In this chapter I do just the opposite: I argue for massive modularity. I then try to show that strong, genetically determined, cognitive predispositions, not only are quite compatible with the kind of cultural diversity we encounter, but even contribute to the explanation of this diversity.

1

How to be a True Materialist in Anthropology

What kind of things are social-cultural things?

(Let me pause immediately. I don't believe that there is any difference between social and cultural things, and I don't want to repeat throughout the phrase 'social-cultural', so I'll toss a coin; head, I will opt for 'social', tail, for 'cultural'. It's tail! From here on 'cultural' means 'social-cultural'. I retain the right, however, to occasionally use 'social' and 'social-cultural', in particular when paraphrasing other people's views.)

What kind of things are cultural things? How do cultural things fit into the world and how do they relate to things other sciences are about? These are philosophical, or, more precisely, ontological questions (ontology, in the classical sense, is the branch of philosophy that tries to answer the question, What is there in the world?, at a *very* abstract level). Ontological questions have practical implications for anthropological research. At stake in particular is the way anthropologists may, or must, collaborate with other disciplines, and the extent to which what they have to say may fit in a general and consistent (though, of course, fragmentary) picture of the world.

This chapter is a synthesis (with revisions) of two earlier articles, 'Issues in the Ontology of Culture', first published in R. B. Marcus et al. (eds), *Logic, Methodology and Philosophy of Science*, vol. 7 (Amsterdam: Elsevier Science Publishers, 1986), 557–71, and 'Les Sciences cognitives, les sciences sociales et le matérialisme', first published in *Le Débat*, 47 (Nov.–Dec. 1987), 105–15. A revised English version of this second article, entitled 'Culture and Matter', came out in J.-C. Gardin and C. S. Peebles (eds), *Representations in Archaeology* (Bloomington: University of Indiana Press, 1991), 56–65.

The natural sciences achieve a high level of mutual consistency and interaction in part because they are all grounded in the same materialist ontology. For a modern materialist,[1] everything that has causal powers owes these powers exclusively to its physical properties. This is taken to be equally true of molecules, stars, rivers, bacteria, animal populations, hearts and brains. Materialism does not imply reductionism. It does not commit scientists who espouse it to describing the objects of their discipline and the causal processes in which these objects enter in the vocabulary of physics. What it does commit them to is describing objects and processes in a manner such that identifying the physical properties involved is ultimately a tractable *problem*, not an unfathomable *mystery* (to use Noam Chomsky's famous distinction – see Chomsky 1975).

Anthropology and Ontology

The world of the social sciences – and in particular that of anthropology – seems, on the other hand, free of any ontological constraint. True, anthropologists occasionally express views about ontology, but these views do not generally carry any methodological commitment. These ontological views are of three types: two types of 'materialism', one empty, the other self-contradictory, and a dualist or pluralist view according to which there is an autonomous cultural level of reality.

The thesis of the ontological autonomy of culture is generally expressed as a series of denials: cultural facts are not biological facts; they are not psychological facts; they are not a sum of individual facts. But what, then, are they? How are they located in space and time? What causal laws do they obey? How do they relate to other kinds of facts? There are no well-argued answers to these questions. The obvious result of assuming that there is a fundamental discontinuity between the biological or the mental on the one side and the cultural on the other is to insulate anthropology from both biology and psychology, and to reject as a priori mistaken any contribution and, even more, any criticism coming from outside. To achieve this dubious outcome, one need not develop in any detail the idea of the autonomy of culture; postulating it is enough.

Empty materialism consists in stating that, of course, everything is material, including social-cultural things, and in leaving it at that. Well and fine; but as long as you do not begin to reflect about the material existence of these things, as long as you keep invoking cause–effect relationships among them without even trying to imagine what material processes might bring about these relationships, you are merely paying lip-service to materialism. You may use for that a number of standard metaphors which evoke the material character of social-cultural things: the mechanical metaphor of social 'forces', the astronomical metaphor of 'revolution', the geological metaphor of 'stratification', and the many biological metaphors of cultural 'life', 'reproduction', and so on. None of these metaphors, however, has ever been developed into a materialistically plausible model. Such empty materialism helps one evade accusations of idealism or of dualism. But otherwise, it remains without effect on one's research practice.

Self-contradictory materialism is a by-product of ill-digested Marxism. It consists of two claims: the first claim is the same as that of empty materialism: everything that is, including social-cultural phenomena, is material. The second is that the material side of the social domain, roughly ecology and economy, determine its non-material side, roughly politics and culture. The contradiction is blatant: the second claim, which contrasts a material and a non-material, or less material, or ideal side of the social domain, is dualistic. It is incompatible, therefore, with materialism in the ontological sense of the term. If materialism is right, then everything is material: law, religion and art no less than forces or relationships of production. From a truly materialist point of view, effects cannot be less material than their causes.

There are two ways to avoid the contradiction of this kind of materialism. The first consists in giving up ontological materialism and in adopting some form of ontological pluralism: there are in the social world both material and non-material things. At this price, it might be conceivable (but don't ask me how) that material things determine non-material ones. The second way out of the contradiction is truer to Marxism (Engels's version at least). It consists in robbing the second claim, that of economic determinism, of any ontological import: one aspect of the material world determines another aspect of the material world. Maybe so, but ontology, and in particular the kind of materialism on which the natural sciences are founded, has nothing to do with it. What remains then, on the ontological side, is the first claim: everything that is, is material; but this is just standard empty materialism.

Looking at the way in which social scientists themselves occasionally articulate their ontological commitments is not the only way – arguably not even the best way – of finding out what these commitments are. It might be more telling to look at social scientists' practice.

Are anthropologists committed to the existence of irreducibly cultural things? Well, they talk with obvious expertise of clans, lineages, marriages, kinship systems, agricultural techniques, myths, rituals, sacrifices, religious institutions, political systems, legal codes and so on. These cultural types are not, and do not correspond to, the types of things talked about in non-social sciences. In particular, they do not correspond to biological or psychological types. Anthropologists have good grounds, therefore, for opposing any kind of reductionism – biological or psychological reductionism in particular – in the study of culture, and good grounds for treating culture as autonomous.

Anthropologists' anti-reductionism, and their commitments to the autonomy of culture, need not, however, be interpreted as true dualism. Anti-reductionism is quite compatible with a modest form of materialism that acknowledges different ontological levels in a wholly material world, as recent developments in the philosophy of psychology show. Could the example of psychology help anthropologists go beyond the unappealing choice between lazy dualism and empty or self-contradictory materialism?

The ontology of psychology: an example to follow?

In psychology too, the choice had long seemed to be between dualism – mental facts and material, in particular neurological, facts are of two radically different natures – and empty materialism. There were materialist metaphors – the biological metaphors of Freud, the mechanical metaphor of Piagetian 'equilibration', for instance – but no materialist model. Only some behaviourists drew practical conclusions from their materialism, but what conclusions! Unable to give a materialist account of mental phenomena, they tried to banish them from the psychological domain!

In the mid-1930s, the mathematician Alan Turing conceived of a materially implementable device that could process information. Even more important, a Turing machine, as the device came to be known, can, demonstrably, perform any operation on encoded information that any other finite physical device, whatever its organization and whatever the way it encodes information, may perform. To put it bluntly, Turing's discovery, and more generally the mathematical theory of automata, seemed to provide a way of understanding how matter can think. It took another twenty years, the development of computers, and important advances in neurology for the impact of Turing's discovery on psychology to be felt, and for a truly materialist approach to cognition to begin emerging.

After decades of behaviourist stricture, mental processes could again be studied, without being reduced to behavioural or neurological processes; but studying them now implied establishing the possibility of their material existence. This in turn implied decomposing them into elementary sub-processes the material implementation of which, for instance on a computer, had become unproblematic.

Establishing the material possibility of a type of mental process – the remembering of a sound pattern, say – is not identical with describing its actual implementation in a human brain. The process of remembering the same information may be performed by means of different types of material processes in different brains, animal or human, or even in the same human brain on different occasions. The process can be programmed in different ways, and differently compiled on different computers. A material model of a type of mental processes may not look at all like the brain processes that actually implement these mental processes. Still, such a model can also serve as a more or less fine-grained hypothesis about the actual material form of the brain processes represented (with testable implications for reaction times, breakdown patterns, etc.).

Whereas stronger versions of materialism imply that psychological *types* can be either reduced to, or eliminated in favour of, neurological types, the more modest materialism involved in current cognitive research does not imply reductionism. It merely implies that *tokens* of mental processes are identical to tokens of neural processes (see Fodor 1974). This modest materialism, when combined with advances in the formal theory of automata and in neurology, is nevertheless a form of true materialism with practical consequences: it imposes strong constraints on acceptable psychological models. It is a materialism with theoretical implications: mental processes are attributed causal powers in virtue of their material properties. Identifying these properties, however difficult, becomes an intelligible task. This modest materialism, nevertheless, grants a modicum of ontological autonomy to the psychological level.

Up to a point, anthropologists might be tempted to follow the example of the cognitive scientists. They might be tempted to assert that, on the one hand, every *token* of a cultural thing is a token of a material thing, while, on the other hand, *types* of cultural things do not correspond or reduce to types of material things. This, however, would be hardly more than an updated reformulation of empty materialism. Why should the anthropological case be different from the psychological one? For two reasons. To begin with, the material locus of psychological processes is obvious enough, and it is homogeneous: token psychological processes are token neurological processes, and the latter have become much better understood. By contrast, if cultural processes have a material realization, then it is a motley one: it involves all kinds of psychological, biological and environmental processes. Secondly, there is no formal discovery akin to Turing's that would provide us with a radically novel insight into the material realization of cultural things. The material character of the types of cultural things recognized by anthropologists remains as mysterious as ever.

There is a deeper disanalogy still between the case of anthropology and that of psychology. The main reason for wanting to adopt modest materialism in psychology is the fact that there is a rich body of accepted knowledge, some of it commonsensical, some of it more sophisticated, the formulation of which requires acknowledging a number of psychological types such as belief, desire, memory, inference, imagination and so on. It seems unwise to forsake this body of knowledge, and unrealistic to see these types as mere terminological instruments without reference in the world. Though not everybody agrees (for different reasons, philosophers such as Churchland

(1988), Dennett (1987) or Stich (1983) do not), psychology as it is seems valuable enough for one to accept its ontological commitments. Modest materialism is a way to combine these commitments with a more general commitment to materialism.

The question now is whether there is in anthropology a body of knowledge worth holding to, and that forces us to acknowledge the existence of irreducible cultural types. David Kaplan argued that such is indeed the case:

> Anthropology has formulated concepts, theoretical entities, laws (or if one prefers generalizations) and theories which do not form any part of the theoretical apparatus of psychology and cannot be reduced to it. This is the logical basis for treating culture as an autonomous sphere of phenomena, explainable in terms of itself. It is wholly beside the point to maintain that anthropologists cannot proceed that way, for the brute fact of the matter is that, in their empirical research, this is the way they do most often proceed. (Kaplan 1965: 973)

Kaplan's argument against reductionism rests, quite appropriately, on an evaluation of anthropology's achievements. This evaluation can be challenged in two ways, one which I will mention but not pursue, since I believe it to be unfair and unproductive, and another which I have developed elsewhere (Sperber 1985b), which, as I will try to show, has some unexpected ontological implications.

The fact is that there is very little agreement among anthropologists about anything, beyond rejecting a few old-fashioned theories, such as meteorological interpretations of religious symbolism, and defending the profession against external attacks. No single concept is shared by all practitioners, and no theory is generally accepted. Under such conditions, it could be argued, nothing can be inferred about the autonomy of culture from the state of the art. I won't pursue this line of argument, because I am convinced that anthropologists, without arriving at any kind of theoretical consensus, have, somehow, developed a genuine, rich, common competence in the study of cultural phenomena. An evaluation of anthropology's achievements which does not include an explication of this competence is incomplete, and therefore insufficient to refute Kaplan's argument.

What I want to suggest, rather, is that what look like 'concepts, theoretical entities, laws and theories' of anthropology are really intellectual tools of another kind. They are interpretive tools. From their existence and usefulness, it is impossible to draw ontological conclusions. One may acknowledge the expertise of anthropologists in matters cultural, and yet deny that they know (or care) what kinds of cultural things really exist. In this respect again, the case of anthropology is quite different from that of psychology.

An interpretive vocabulary

It is not so much that anthropologists don't *share* theoretical concepts; it is that they don't *have* theoretical concepts of their own. What they do have is a collection of technical terms – technical in the sense that they are terms of the trade, rather than ordinary language terms (or they are ordinary language terms used in a non-ordinary way). They are not theoretical, though, in that their origin, development, meaning and use are largely independent of the development or content of any genuine theory (of the kind that carries ontological commitments).

Throughout the history of anthropology, a number of technical terms have been critically analysed: for instance, 'taboo' by Franz Steiner (1956) and Mary Douglas (1966), 'totemism' by Goldenweiser (1910) and Lévi-Strauss (1966b), 'patri-' and 'matrilinearity' by Leach (1961), 'belief' by Needham (1972), and, of course, 'culture' by a great many anthropologists (see Kroeber and Kluckhohn 1952, Gamst and Norbeck 1976). The vagueness or the arbitrariness of these terms has been repeatedly pointed out. Yet, in spite of this critical work, there are no signs that anthropologists are converging on a set of progressively better defined, better motivated notions. If anything, there is more divergence and no greater conceptual precision today than there was half a century ago.

Leach (1961) and Needham (1971, 1972, 1975) have convincingly argued that this vagueness of anthropological terms is not accidental, that it has to do with the way these terms have developed and with the kinds of things they are being used to refer to; so, if we want proper theoretical terms in anthropology, we should construct altogether new ones.

Needham has further argued that anthropological technical terms are best understood as 'family resemblance' or 'polythetic' terms – that is, as terms referring to things among which resemblances exist, but which don't fall under a single definition. The classical example, developed by Wittgenstein (1953), of a 'family resemblance' term is that of 'game'. Among the characteristic features of games are 'competitive', 'amusing' or 'obeying rules'. However, patience is not competitive; playing chess need not be an amusement; and one may play with a ball without definite rules. More technically, a polythetic term is characterized by a set of features such that none of them is necessary, but any large enough subset of them is sufficient for something to fall under the term.

A polythetic term need not be fully polythetic: all its referents may share one or even several features; but as long as these necessary features are not jointly sufficient, the term is still polythetic. Actually, it is doubtful that fully polythetic terms (i.e. terms without any necessary feature) are ever used. All members of a useful polythetic class normally belong to the same domain, which determines at least one common feature. All the members of the class of 'games' share the feature of being activities.

I want to argue that anthropological terms do indeed have some kind of family resemblance organization, but that it is a different kind of family resemblance from the one Wittgenstein and Needham had in mind. They had in mind a resemblance between the things described by the same term. For instance, every thing described as a game resembles some other things described as games. Let us call this a 'descriptive resemblance'.

Anthropological technical terms are not used simply to describe, however; they are used to translate or render native terms or notions (or notions that the anthropologist attributes to the natives). Anthropological terms are used not descriptively, but interpretively. The resemblance involved is a resemblance in meaning among all the notions rendered by means of the terms, rather than a resemblance among the things referred to by these terms. It is what we may call an 'interpretive resemblance'. All the notions that can be properly interpreted by means of the same term will exhibit a typical family resemblance pattern: even though it may happen that two such notions fail to resemble one another, there is sure to be at least one further notion that they both resemble.

You might be tempted to say: here is a distinction without a difference; for, after all, to the extent that terms have similar meanings, then the things they denote are similar too. But terms can have meaning without having denotation – without, that is, referring to anything. Thus the resemblance between 'goblin', 'leprechaun', 'imp', and 'gnome' is a resemblance among meanings, among ideas, and not among things. If an anthropologist were to use the word 'goblin' to render a notion of the people she studied, she would not commit herself to the existence of goblin-like creatures, but only to the existence of 'goblin'-like representations. Interpretively used terms do not commit their user to the existence of the things purportedly denoted by these terms.

The view that anthropology is essentially an interpretive science is well known. It has been famously defended by Clifford Geertz (1973). I agree that anthropologists studying individual cultures are – quite properly, too – mainly involved in an interpretive task; that is, in representing native representations by means of translations, paraphrases, summaries and syntheses understandable to their readers. I see nothing objectionable in the fact that, for this purpose, they should use an interpretive vocabulary. However, if I am right in claiming that the anthropological vocabulary is interpretive, then anthropological accounts are wonderfully free of ontological commitments. Just as the appropriate use of 'goblin' by an anthropologist tells us nothing regarding the existence of goblins, the appropriate use of 'marriage', 'sacrifice', or 'chiefship' does not tell us whether marriages, sacrifices or chiefships are part of the furniture of the world.

'Marriage'

Take 'marriage'. Here is a true technical term of anthropology, and as good a type of cultural thing as you will ever get. But how good a *type* is it? Do all marriages fall under a single definition, or do we have reasons otherwise to believe that they share some yet unanalysed common essence?

Let us look first at a couple of characterizations of marriage that have been proposed. *Notes and Queries* (1951) suggested: 'Marriage is a union between a man and a woman such that children born to the woman are recognized legitimate offspring of both partners.' Here, you don't have to look for exotic counter-examples. In most Western societies, the distinction between legitimate and illegitimate offspring is becoming abolished. Children born in or out of wedlock may enjoy the same rights. The only sense in which some children may still be called 'illegitimate' is precisely that they are born out of wedlock. But this, of course, makes a definition of marriage in terms of the legitimacy of offspring quite circular.

Or consider Lévi-Strauss's claim: 'If there are many types of marriage to be observed in human societies ... the striking fact is that everywhere a distinction exists between marriage, i.e. a *legal, group-sanctioned bond between a man and a woman,* and the type of permanent or temporary union resulting either from violence or consent alone' (1956: 268, my emphasis). In the very same paper, Lévi-Strauss gives a counter-example to his own characterization. He argues that many 'so-called polygamous societies ... make a strong difference between the 'first' wife who is the only true one, endowed with the full rights attached to the marital status while the other ones are sometimes little more than official concubines' (ibid. 267). The bond between a man and his official concubine is, surely, group-sanctioned, or in what sense is it 'official'? Therefore, if Lévi-Strauss wants to distinguish this bond from true marriage, his characterization of marriage fails.

Such failures to properly define 'marriage' are not accidental. Leach has argued that 'marriage is ... "a bundle of rights"; hence all universal definitions of marriage are vain' (1961: 105). The relevant rights, he argues, vary from society to society. Leach lists ten kinds of rights, from 'to establish the legal father of a woman's children' to 'to establish a socially significant "relationship of affinity" between the husband and his wife's brothers'. He shows that there is not a single one of these rights which is present in all cases of marriage.

Pushing further Leach's argument, Needham concludes that 'marriage' 'is an odd-job word: very handy in all sorts of descriptive sentences, but worse than misleading in comparison and of no real use at all in analysis' (1971: 8). There are two ways in which the word 'marriage' can be said to do odd jobs: it does a few different jobs for all anthropologists; in addition and more importantly, it does a different job for each anthropologist in his or her own field.

Imagine an anthropologist who goes to study the Ebelo. She might, in principle, wonder whether the Ebelo have the institution of marriage at all, but it would be surprising if she did. It is generally taken for granted among anthropologists that marriage is universal. Our anthropologist does not, however, expect to come across some practice that would fall squarely under a well-established definition of marriage, if only because there is no such definition. What she expects to find is some native institution which she may call 'marriage' with as much justification as other anthropologists in their use of the word.

The problem she faces is not whether the Ebelo have marriage, but, as Peter Rivière puts it, 'which of the forms of relationship between the sexes is ... to be regarded as the marital one' (1971: 65). The logic is one of a party game: 'if one of these forms of relationship were one of marriage, which one would it be?' It would take a very odd society, or a very uncooperative anthropologist, for the question to remain without an answer. It is not surprising, then, that marriage should be found in every society. This is made possible, however, precisely by the fact that 'marriage', whatever it does otherwise, does not denote a precise type of cultural thing.

How, then, does our anthropologist go about identifying which Ebelo form of relationship is 'the marital one'? Does she *look* at relationships? No, relationships are not the kind of things you can look at. What she does, mostly, is to get Ebelo people to describe in their own terms the types of relationships they entertain among themselves. She then decides which of the native notions and, possibly which of the native terms is best rendered by 'marriage'.

Our anthropologist comes to the conclusion that 'marriage' corresponds to the Ebelo term *kwiss*. She now explains what she understands the Ebelo to believe: namely, that marriage — that is, *kwiss* — is a bond between a man and a woman blessed by ancestral spirits. Note in passing that 'bond', 'blessed', 'ancestral spirit' are also used interpretively in this characterization of the meaning of 'marriage'/*kwiss*. That is, they are not used to describe things, but to render yet other Ebelo notions.

A new case of marriage, the Ebelo case, has now been added to the anthropological stock. It has been added on the basis of a resemblance. So were all previous cases, even the first one. 'Marriage' became a technical term of anthropology when an anthropologist – or was it a historian? – decided that some exotic or ancient notion was best rendered by the ordinary language word 'marriage'. From then on 'marriage' began to swell and loose its contours as more and more different notions were interpreted by means of it. The meaning of 'marriage' in anthropological writings became a loose synthesis or compound of the sundry particular notions the term served to interpret. The point to stress is that for a new notion to be rendered by 'marriage', it need not fall under any general notion conveyed by the term; all it need do is resemble in content the bloated meaning of the anthropological term. That is why the fuzziness of anthropological terms is no obstacle to their use. Fuzziness is never a hindrance – if anything, it is a help – to the establishment of resemblances.

That the anthropological notion of marriage should be a family resemblance notion is no accident. It is a result of the very way in which it has been, and is being, developed. Resemblance – not the possession of definite features – determines where 'marriage' is to be applied. There is no reason to expect the development of anthropology to reverse this state of affairs. Actually, the better anthropologists come to know a greater variety of cases, the looser becomes the resemblance between instances of marriage.

But the resemblance involved in determining the applicability of 'marriage' one among the things called marriage or one between the notion interpreted and the notion (or notions) generally conveyed by the term used to interpret it? Is it, in other words, a descriptive or an interpretive resemblance? If the account I have sketched of how anthropologists go about identifying new cases of marriage is correct, then clearly the resemblance involved is an interpretive one.

Implications

The two types of family resemblance, the descriptive and the interpretive, carry different ontological implications. If you take 'marriage' to be based on descriptive resemblance, you should envisage that the term is only partly polythetic. Surely, all marriages are social ties; they all involve legal rights and duties. So when you describe something as a marriage, this may well commit you to the existence of social ties and of legal rights and duties as basic types of things in the domain of the social sciences. These sociological types do not seem to reduce to types of things recognized in neighbouring sciences such as psychology, biology, or ecology. Hence typologies based on descriptive family resemblance, in spite of all their fuzziness, may nevertheless commit you to the existence of irreducible cultural types.

Not so with interpretive resemblance. Imagine that our anthropologist reports that two Ebelo individuals, say Peter and Mary, are married. Is she, in so doing, *stating* that there is a bond between Peter and Mary that has been blessed by ancestral spirits? Presumably not, if only because it would commit her to the existence of ancestral spirits. She is reporting, rather (in free indirect style – see Sperber 1985b) what the Ebelo people involved believe about Peter and Mary. She is interpreting Ebelo ideas. What does such an interpretation commit her to, ontologically speaking? It commits her to the existence of certain Ebelo people and to the existence of certain representations in the minds of these people. Does it commit her to the existence of a thing or a state of affairs which could appropriately be called a marriage? I don't see how. Our anthropologist, if she is like other anthropologists, may be disposed to further commit herself anyhow. She probably takes it for granted that there is such a thing as marriage, instantiated, in particular, among the Ebelo. But nothing in her account of the Ebelo *kwiss* would compel us to follow her in such a commitment. All we know for sure, if we believe her ethnography, is that some Ebelo people have formed the belief that Peter and Mary (and so many others) are *kwiss*ed. But why should we share this belief, or even a rationalized version of it in which ancestral spirits are omitted?

But what about the use of the term 'marriage' in theoretical or comparative anthropological work? Doesn't it, there at least, correspond to a general concept? Well, if you believe it does, say *which* concept. I am not claiming that it would be impossible to define a general concept which could reasonably be expressed by 'marriage'.[2] I am merely suggesting that there is no obvious reason why you would want to define a concept meeting this particular condition, and that anthropologists, notwithstanding the appearances, have never truly bothered. They have found it useful to abstract from

interpretive ethnographic reports in order to arrive at general interpretive models. These models are not *true* of anything; what they do
is help the reader to get a synthetic view of ethnographic knowledge. They also serve as sketches of possible interpretations for further ethnographic work. So, 'marriage' in these general
anthropological writings is both a loose, topic-indicating word and
an interpretive term used synthetically.

What is true of 'marriage' is true of the technical vocabulary of
anthropology in general. 'Tribe', 'caste', 'clan', 'slavery', 'chiefship',
'state', 'war', 'ritual', 'religion', 'taboo', 'magic', 'witchcraft', 'possession', 'myth', 'tales', and so on are all interpretive terms. There is
a family resemblance – an interpretive one – between all the
notions each of these terms serves to render. When these terms are
used to report specific instances of events or states of affairs, they
help the reader to get an idea of the way in which the people concerned perceive the situation ('seeing things from the native's point
of view', as the phrase goes). What do these interpretive reports tell
us about the nature of whatever *is* taking place? Well, what they
tell us for sure is that some representations are being entertained
and communicated.

A few general terms used in anthropology are not interpretive in
that sense, but nor do they suggest the existence of a distinct ontological level of culture. Some are straightforwardly psychological,
such as 'colour classification'. Others are straightforwardly ecological, such as 'sex ratio'. What differentiates these psychological or
ecological terms used in anthropology from the proprietary vocabulary of the field is that they apply quite independently of the 'point
of view' of the people concerned. People can have a colour classification without being aware of the existence of such things as classifications; and not only human, but also other animal populations can
have sex ratios without having thoughts of sex ratio. On the other
hand, even an anthropologist wholly committed to the existence of
marriages would agree that no actual marriage is ever entered into
unless the appropriate people entertain the idea that a marriage (or a
kwiss, or something of the sort) has been entered into. Again, the
only sure thing when a man and a woman are said to be married is
that some representations of their being married, or *kwiss*ed, are
being circulated.

What Are Cultural Things Made Of?

Let us start as simply as possible. Cultural things are, in part, made of bodily movements of individuals and of environmental changes resulting from these movements. For instance, people are beating drums, or erecting a building, or slaughtering an animal. The material character of these phenomena is, so far, unproblematic. But we must go further. Is it a musical exercise, a drummed message, or a ritual? Is it a house, a shop, or a temple? Is it butchery, or sacrifice? In order to answer, one must, one way or another, take into account the representations involved in these behaviours. Whatever one's theoretical or methodological framework, representations play an essential role in defining cultural phenomena. But what are representations made of?

Let us note, to begin with, that two types of representations are involved: mental representations and public representations. Beliefs, intentions and preferences are mental representations. Until the cognitive revolution, the ontological status of mental representations was obscure. Signals, utterances, texts and pictures are all public representations. Public representations have an obviously material aspect. However, describing this aspect – the sounds of speech, the shapes and colours of a picture – leaves out the most important fact, that these material traces can be interpreted: they represent something for someone.

In order to account for the fact that public representations are interpretable, one must assume the existence of an underlying system: for example, a language, a code, or an ideology. In the semiotic and semiological traditions, these underlying interpretation systems have been described in abstract, rather than in psychological terms, and indeed their existence has often been considered extra-psychological. With such an approach, the material existence of these systems remains obscure. As a result, the material properties which make public representations interpretable and the material existence of cultural phenomena described with reference to public representations remain obscure too. One may also view underlying interpretation systems as complex mental representations; this is, for instance, what Noam Chomsky does when he describes a grammar as a mental device. This second approach brings us back to the psychology of mental representations, and therefore to the new perspectives opened by the development of the cognitive sciences.

The biggest difficulty in developing even a minimal materialism in the social sciences came from the role that representations unavoidably play in them. However, in psychology, the material character of mental representations has changed from the status of a mystery to that of an intelligible problem. The question is whether the social sciences can redefine their notion of a representation on the basis of the cognitive notion of a representation. What I would like to do now is to suggest how this can be done, and how, as a result, the whole ontology of the social sciences can be tightened, how a truly materialist programme in the social sciences becomes conceivable.

An epidemiology of representations

Just as one can say that a human population is inhabited by a much larger population of viruses, so one can say that it is inhabited by a much larger population of mental representations. Most of these representations are found in only one individual. Some, however, get communicated: that is, first transformed by the communicator into public representations, and then re-transformed by the audience into mental representations. A very small proportion of these communicated representations get communicated repeatedly. Through communication (or, in other cases, through imitation), some representations spread out in a human population, and may end up being instantiated in every member of the population for several generations. Such widespread and enduring representations are paradigmatic cases of cultural representations.

The question is: Why do some representations propagate, either generally or in specific contexts? To answer such a question is to develop a kind of 'epidemiology of representations'. The epidemiological metaphor can help us, provided we recognize its limits. One limit is self-evident: we certainly do not want to imply that cultural representations are in any sense pathological. Another limit, though less obvious, is much more important: whereas pathogenic agents such as viruses and bacteria reproduce in the process of transmission and undergo a mutation only occasionally, representations are transformed almost every time they are transmitted, and remain stable only in certain limiting cases. A cultural representation in particular is made up of many versions, mental and public ones. Each mental version results from the interpretation of a public representation which is itself an expression of a mental representation.

One might choose as a topic of study these causal chains made up of mental and public representations, and try to explain how the mental states of human organisms may cause them to modify their environment, in particular by producing signs, and how such modifications of their environment may cause a modification of the mental states of other human organisms. (What I call 'chains' are, of course, quite complex, and generally look like webs, networks, or lattices. Still, they are all made of only two types of links: from the mental to the public and from the public to the mental.) The ontology of such an undertaking resembles that of epidemiology. It is a rather heterogeneous ontology, in that psychological and ecological phenomena are mixed together, just as in epidemiology, pathological and ecological phenomena are mixed. In each case, what is to be explained is the distribution of individual conditions, pathological or psychological. And in each case the explanation takes into account both the state of the individuals and that of their common environment, which is itself largely modified by the behaviour of the individuals.

In spite of its heterogeneity, the ontology of an epidemiology of representations is strictly materialist: mental representations are brain states described in functional terms, and it is the material interaction between brains, organisms and environment which explains the distribution of these representations.

Because of the ontological heterogeneity of epidemiological phenomena, there is no such thing as a general epidemiological theory. What we find is a variety of different models with greater or lesser generality and a common methodology. Similarly, I very much doubt that we should, in the study of cultural phenomena, aim at a grand general theory.

Different types of representations may have their distribution explained in quite different ways. For the time being, at least, a realistic and ambitious enough aim would be to develop materialistically plausible explanatory models of the distribution of, for instance, various folk classifications, myths, techniques, art-forms, rituals, legal rules and so forth. Models with the greatest possible generality, provided they are truly explanatory, are of course to be preferred. However, aiming from the start at a holistic theory, as many social scientists are prone to do, results, for practical – and possibly also for substantive – reasons, in no theory at all.

Let me illustrate very briefly the epidemiological approach with a couple of examples.

'Myth'

Take a myth, say the Bororo myth of the bird-nester which Lévi-Strauss uses as the starting-point of his *Mythologiques*. In a traditional approach, this myth would be presented in the form of a canonical version arrived at by selectively synthesizing the various versions collected. Such a canonical version is an abstract object, without existence in the society studied. It may serve an expository purpose, but, as it stands, it neither calls for, nor provides, an explanation. Lévi-Strauss himself departs from this traditional approach: for him, to study a myth is to study the relationships of 'transformation' (i.e. the way in which resemblances and differences are patterned) between the different versions of the myth and between the myth and other myths. With this approach, neither a single version nor a synthesis of several versions is an appropriate object of study. A myth should be considered, rather, as the set of all its versions.

The ontological status of a myth as a set of versions and the explanatory value of analysing the relationships of transformations between these versions are unclear, but they can be clarified in an epidemiological perspective. What I propose, in a nutshell, is to try to model not the set, but the causal chains linking the different versions of the myth, and this entails considering not just the public versions, but also the mental ones (without which there would be no causal chain). Of course, we have records of only a few of the public versions and none of the mental ones, but complementing observations with hypotheses about unobserved – and even unobservable – entities is plain normal science. Moreover, in order to explain reasonably well the distribution of versions of the same myth, our task would not be to describe all the links in the chain – that is, all the individual steps of mental transformation and public transmission. It would be to give an account of the type of causal factors that have favoured transmission in certain circumstances and transformation in certain directions.

Studying a myth from this perspective, we have three types of objects:

1   Narratives: that is, public representations which can be observed and recorded, but which can only be interpreted by taking into account:
2   Stories: that is, mental representations of events, which can be expressed as, or constructed from narratives.
3   Causal chains: stories – narratives – stories – narratives . . .

Every token of one of these three types is a material object. Every token of a narrative is a specific acoustic event. Every token of a story is a specific brain state. A chain causally linking such specific material things is, of course, itself a material thing.

The causal explanation of the existence of these public narratives and these mental stories is provided by the description of the causal chains in which they occurred. The explanation of such a causal chain calls for a model where in both psychological and ecological factors come into play. For instance, a crucial ecological factor would be the absence in the society considered of the kind of external memory stores which writing provides; oral representations, unlike written ones, are environmental events rather than environmental states. A crucial psychological factor would be the organization of spontaneous human memory. The interaction of these two factors would help explain why a given narrative with an easily memorized structure is transmitted with little variation in an oral tradition.

What about the old anthropological concept of a myth in all this? A chain of versions is, of course, no more a myth than an epidemic of influenza is a case of influenza. Unlike a case of influenza, which is influenza even without an epidemic, each mental story or each public narrative is itself cultural, and hence mythical, only to the extent that it belongs to such a chain. No material object, therefore, is intrinsically a myth. At best, talking of myth may serve to draw attention to a body of related data. But the basic concept needed in studying these data is that of a causal chain of narratives and stories. A truly materialist ontology leads to a reconceptualization of the domain.

The fact that in a non-literate population we find narratives that can be considered versions of one another (and are so considered by the natives) is what leads us to identify a 'myth'. Other cultural forms, such as 'beliefs', 'folk classifications', 'traditional techniques', are also characterized by a wide distribution of very similar representations. The ontological constraints on the concepts involved in the study of such phenomena are clear enough. They call for eliminating abstract synthetic versions of these representations and keeping only the many public and mental versions and their causal chains.

'Marriage' again

The ontology of social institutions, which are the subject-matter *par excellence* of the social sciences, raises further problems. For there to be a state, a market, a church, a ritual, it is not necessary that every individual who participates in the institution should have a mental version of it; indeed, in most cases the very idea is meaningless. Institutions are neither public nor mental representations. How, then, could an epidemiology of representations help provide a materialist account of institutions?

Well, an epidemiology of representations is not about representations, but about the process of their distribution. In some cases, similar representations – for example, versions of the same myth – are distributed by a repetitive chain of public and mental representations; in other cases, many different representations, the contents of which do not at all resemble one another, are involved in the same distribution process. In particular, some of the representations involved may play a regulatory role by representing how some of the other representations involved are to be distributed. The distribution of these regulatory representations plays a causal role in the distribution of the other representations in the same complex. Institutional phenomena, I maintain, are characterized by such hierarchical causal chains.

I again take the example of 'marriage', but in a version better known than the Ebelo one: namely, civil marriage in France today.

The classical approach would consist in defining marriage as a jural link of a certain type between a woman and a man established through a specific ritual, the marriage ceremony. If natives of France today, the author of this book included, have no problem using these notions, we should be puzzled as scientists, by the ontological status of a ceremony, of a jural link, and therefore of marriage itself. The epidemiological approach provides a solution to the puzzle.

The material process which causes the natives to say, for instance, that Pierre and Marie got married involves two levels of representation. At the higher level, there is a regulatory representation of a course of action: certain pre-conditions being fulfilled, a civil officer 'pronounces a man and a woman united by marriage'. The basic public version of this representation is a chapter of the *Code civil*, the origin and distribution of which is, to a large extent, a matter of public record. Note that what this higher-level representation describes is a type of lower-level representation and the conditions under which versions of it can be produced and distributed. The civil officer who pronounces Pierre and Marie husband and wife produces such a lower-level representation in accordance with the higher level, regulatory one. The lower-level representation can then be reproduced, paraphrased, elaborated and so on. Anyone who now says that Pierre and Marie are married is not describing an actual material fact, but restating the civil officer's original representation. Anyone who states the rights and duties of Pierre and Marie as married people is producing, with reference to the case of these two individuals, a more or less faithful version of a higher-level, general representation of the rights and duties of married people, the original version of which is again in the *Code civil*.

'Marriage', 'rights', 'duties', are immaterial entities existing in the ontology of the natives, hence in our everyday ontology. In our materialist ontology for scientific use, on the other hand, there exist only mental or public *representations of* marriage in general, of particular marriages, and of rights and duties, and the complex causal chain in which these representations occur. The representations which natives have of immaterial entities are themselves quite material. The distribution of these representations can have effects on the behaviour of the natives which are very similar to the effects that the natives themselves attribute – wrongly – to the immaterial state

of affairs represented. The difference of ontologies is not incompatible, therefore, with a certain degree of correspondence between the
two descriptions, that of the native and that of the scientist.

The programme I am suggesting is not without precursors. The dif-
fusionist approach in anthropology and archaeology was concerned
with a causal explanation of the distribution in space and time of
cultural items. One of its weaknesses was the poverty of its psychological assumptions. A comparable weakness is found in several
recent biologically inspired approaches to culture, where the
mind/brain is essentially seen as a duplication device (see e.g. Boyd
and Richerson 1985, Cavalli-Sforza and Feldman 1981, Dawkins
1976, Lumsden and Wilson 1981). The most obvious lesson of
recent cognitive work is that recall is not storage in reverse, and
comprehension is not expression in reverse. Memory and communication transform information. Thus, to treat representations,
whether mental or public, as material causes among other material
causes implies rooting the study of thought and of communication
in cognitive psychology.

An epidemiology of representations would establish a relationship
of mutual relevance between the cognitive and the social sciences,
similar to that between pathology and epidemiology. This relationship would in no way be one of reduction of the social to the psychological. Social-cultural phenomena are, on this approach,
ecological patterns of psychological phenomena. Sociological facts
are defined in terms of psychological facts, but do not reduce to
them.

2

Interpreting and Explaining Cultural Representations

A representation sets up a relationship between at least three terms: that which represents, that which is represented, and the user of the representation. A fourth term may be added when there is a producer of the representation distinct from its user. A representation may exist inside its user: it is then a *mental representation*, such as a memory, a belief, or an intention. The producer and the user of a mental representation are one and the same person. A representation may also exist in the environment of its user, as is the case, for instance, of the text you are presently reading; it is then a *public representation*. Public representations are usually means of communication between a user and a producer distinct from one another.

A mental representation has, of course, a single user. A public representation may have several. A speech may be addressed to a group of people. A printed text is aimed at a wide audience. Before techniques such as printing or magnetic recording made the strict duplication of a public representation possible, oral transmission allowed the production of representations similar to one another: the hearer of a tale may, for instance, become in turn its teller. It must be stressed, however, that oral transmission is not a reliable means of *reproduction*; it generates a fuzzy set of representations which are more or less faithful *versions*, rather than exact copies, of one another.

This chapter is a revision of an article with the same title first published in G. Palsson (ed.), *Beyond Boundaries: Understanding, Translation and Anthropological Discourse* (Oxford: Berg, 1993), 162–83. A shortened version had been published in French as 'L'Étude anthropologique des représentations: problèmes et perspectives', in D. Jodelet (ed.) *Les Représentations sociales* (Paris: Presses Universitaires de France, 1989), 115–30.

Consider a social group: a tribe, the inhabitants of a town, or the members of an association. Such a group and its common environment are, so to speak, inhabited by a much larger population of representations, mental and public. Each member of the group has, in his or her head, millions of mental representations, some short-lived, others stored in long-term memory and constituting the individual's 'knowledge'. Of these mental representations, some – a very small proportion – get communicated repeatedly, and end up being distributed throughout the group, and thus have a mental version in most of its members. When we speak of *cultural representations*, we have in mind – or should have in mind – such widely distributed, lasting representations. Cultural representations so understood are a fuzzy subset of the set of mental and public representations inhabiting a given social group.

Anthropologists have not converged on a common view of cultural representations, a common set of questions about them, or even a common terminology to describe them. Most authors approach the various genres of representation separately, and talk of beliefs, norms, techniques, myths, classifications and so forth according to the case. I would like, nevertheless, to reflect on the way in which anthropologists (and other social scientists) represent and attempt to explain cultural representations in general.

Interpreting Cultural Representations

Suppose you want to produce a representation of a basket: you may produce an image of the basket, or you may describe it. In other words, you may either produce an object that resembles the basket – for instance, a photograph or a sketch – or you may produce a statement. The statement in no way resembles the basket, but it says something true about it. (Truth, of course, is a necessary, but not a sufficient, condition for a description to be adequate.) It might seem that the situation is the same when what you want to represent happens to be a representation: the tale of 'Little Red Riding Hood', for instance. You might record or transcribe the tale (or, rather, a particular version of it): that is, produce an object that resembles the tale in the manner in which a photograph or a sketch resembles a basket. You might also describe the tale by saying, for instance: 'It is a tale found throughout Europe, with one animal and several human characters.'

Yet, there would be something missing in these representations of 'Little Red Riding Hood': the recording or the transcription in itself represents only an acoustic form, while the description suggested tells us little more about the *content* of the tale, which, after all, *is* the tale. All you need do, it might be argued, is describe the tale in greater detail. You might say for instance: ' "Little Red Riding Hood" is a tale found throughout Europe, which tells the story of a little girl sent by her mother to take a basket of provisions to her grandmother. On her way, she meets . . .' You could, of course, in this manner recapture the content of the tale as closely as you wish; but notice what would be happening: instead of *describing* the tale, you would be *telling* it all over again. You would be producing an object that represents the tale, not by saying something true about it, but by resembling it: in other words, you would be producing yet another version of the tale.

Let us generalize: in order to represent the content of a representation, we use another representation with a similar content. We don't describe the content of a representation; we paraphrase it, translate it, summarize it, expand on it – in a nutshell, we *interpret* it.[3] An *interpretation* is a representation of a representation by virtue of a similarity of content. In this sense, a public representation, the content of which resembles that of the mental representation it serves to communicate, is an interpretation of that mental representation. Conversely, the mental representation resulting from the comprehension of a public representation is an interpretation of it. The process of communication can be factored into two processes of interpretation: one from the mental to the public, the other from the public to the mental.

Interpretations are just as ordinary in our mental life as are descriptions; they are a form of representation produced and understood by everyone. To express oneself or to understand other people's expressions is, implicitly, an act of interpretation. Moreover, we are all producing explicit interpretations when answering questions such as, What did he say? What does she think? What do they want? In order to answer such questions, we represent the content of utterances, thoughts, or intentions by means of utterances of similar content.

Needless to say, the anthropological study of cultural representations cannot ignore their contents. As a result, and whether this pleases us or not, the task of anthropology is in large part interpretive. However, precisely because interpretation is based on a quite ordinary ability, rather than on a sophisticated professional skill, most anthropologists have produced interpretations just as Molière's Monsieur Jourdain produced prose: without being aware of doing so, or at least without reflecting much on the matter.

As long as interpretation is about individual words or thoughts, the degree of freedom that the interpreter grants herself may be manifest and unproblematic. You inform me, for instance, in one sentence and with a sneer, of what the Prime Minister said in his press conference; I have no trouble understanding that while the gist may be the Prime Minister's, the conciseness and the irony are yours. Similarly, ordinary anthropological renderings *of individual words and thoughts* are, often enough, easy to understand and to accept (as long as not too much depends on exact formulation, and, even then, a carefully glossed translation may still do).

In anthropology, however, what gets interpreted is often a *collective representation* attributed to a whole social group ('The So-and-so believe that . . .'), one which need never be entertained, let alone expressed, by any one individual member of that group. There is neither a clear common-sense understanding of what such a collective representation might be, nor a straightforward way of assessing the faithfulness of the rendering. The lack of a clear methodology makes it difficult to evaluate, and hence to exploit, these interpretations. Nevertheless, they are given an important role in anthropological accounts, and, as we will see, are even offered as ultimate explanations sometimes.

Here is an illustration. The scene, reported by the French anthropologist Patrick Menget, takes place among the Txikao of Brazil.

At the end of a rainy afternoon, Opote came back home carrying a fine *matrinchao* fish he had caught in his nets. He put it down without a word next to Tubia, one of the four family heads of his house. Tubia cleaned it and put it on to smoke. Until the fall of night he ate it, by himself, in small mouthfuls, under the interested eyes of the other inhabitants of the house. No one else touched the *matrinchao*,

nor showed any desire to have some of it. Yet the hunger was universal, and the flesh of the *matrinchao* is among the most highly
praised. (Menget 1982: 193)

So far, this is essentially an ordinary description: every sentence in
it expresses a proposition presented by the anthropologist as true.
The situation described, however, is quite puzzling: 'Why', asks
Menget, 'this general abstention?' And he goes on to answer:

> The fisherman, Opote, possessor of fishing magic, could not con-
> sume his catch without the risk of damaging this magic. The other
> family heads avoided the flesh of the *matrinchao* for fear of endanger-
> ing the health and the lives of their young children, or their own
> health. Since their wives were nursing, they had to abstain for the
> same reason. The children, finally, would have absorbed the particu-
> larly dangerous spirit of this species. (Ibid.)

This time the anthropologist – who does not himself believe in
magic or spirits – is not presenting as true that Opote was running
the risk of damaging his magic, or that the children would have
absorbed a particularly dangerous spirit. He is presenting these state-
ments as similar in content to the beliefs motivating the abstinence
of Opote's people. These are interpretations. Such interpretations of
individual thoughts are neither harder to comprehend nor more sus-
picious than the interpretations we all use all the time to talk of each
other.

However, the anthropologist's ultimate goal is not to report particular events. Menget's aim, for instance, in reporting the anecdote
of Opote's *matrinchao*, was to illustrate some hypotheses on the 'couvade', first among the Txikao themselves, then among the South
American Indians, and ultimately on the couvade in general.
'Couvade', it will be remembered, refers in anthropological literature to a set of precautions (e.g. resting, lying down, food restric-
tions) a man is expected to take during and just after the birth of a
child of his, precautions similar to those imposed, more understand-
ably, on the mother of the child (see Rivière 1974).

Menget proposes a subtle analysis of the relevant Txikao views on
life and its transmission, and concludes:

[E]verything happens as if two antagonistic principles ruled over the life processes. . . . A strong principle, tied to blood, to fat, rich meats and fermentation results from the constant somatic transformation of weaker substances, water, milk, sperm, white flours, lean meats. But inversely, the human body, in rhythms that vary with age, sex, and condition, anabolizes the strong substances and neutralises their danger.

. . . In the couvade, the whole set of occupational, alimentary, and sexual taboos comes down in the end to avoiding either an excess of strong substances . . . or a loss of the weak somatised substances. . . . The creation of a new human being activates the whole universal process of transformation of substance, but also the separation of a part of the somatised substance of the parents and the initiation of an individual cycle. (Menget 1982: 202–03)

Again, the ethnographer is interpreting: he does not himself, for instance, believe or intend to assert that 'the human body anabolizes strong substances', the non-assimilation of which 'leads to swelling diseases'. He is offering such formulations as similar in content to cultural representations underlying the Txikao couvade practices.

However, while it is easy enough to imagine Opote thinking or saying, in roughly similar terms, that he could not consume his catch without damaging his magic, it is hard to conceive of Txikao thoughts or utterances involving notions of, say, the 'somatic transformation of weak substances' or the 'anabolization of strong substances'. The resemblance in content between the interpretation and the representations interpreted is manifestly weaker here than in ordinary interpretations of individual thoughts or utterances, and the degree of resemblance is hard or even impossible to evaluate. (What is at stake is not the work of an individual anthropologist: on the contrary, I have chosen to discuss Menget's essay because I see it as a good example of today's best ethnography. At stake here are the limits inherent in the interpretive approach to cultural representations.)

An ethnographer is faced at first with a great diversity of behaviours which he progressively comes to understand by discerning underlying intentions: that is, by becoming able to conceptualize these behaviours as actions. He becomes adept, in particular, at discerning the intentions governing speech acts, or, in other terms, at comprehending what his interlocutors mean.

Intentions thus understood still call for further, deeper understanding. Let us accept that 'the other family heads avoided the flesh
of the *matrinchao* for fear of endangering the health and the lives of
their young children, or their own health.' But how are such means
supposed to serve such ends? A deeper understanding of intentions
involves grasping how they could be rational, or, in other words,
seeing how they might follow from underlying desires and beliefs.
If, for the Txikao, the flesh of the *matrinchao* is 'strong' and haz-
ardous for one's health, if father and child are of one and the same
substance, a substance which, contrary to appearances, does not
divide into two independent beings until some time after birth, then
we begin to grasp how the behaviour of Opote's people might be
rational. To grasp it further, we should try to establish the rationality
of the underlying beliefs themselves – that is, not just their mutual
consistency, but also their compatibility with Txikao experience.

In our everyday striving to understand others, we make do with
partial and speculative interpretations (the more different from us the
others, the more speculative the interpretations). For all their incom-
pleteness and uncertainty, these interpretations help us – us individuals, us peoples – to live with one another. Anthropologists have
contributed to a better understanding, and thus a greater tolerance, of
culturally different others. To do so, they haven't relied on scientific
theories or rigorous methods, which are not part of the anthropologist's standard tool kit. Given the cultural distance, the comprehension goals of anthropologists are particularly ambitious and arduous.
Still, the form of comprehension involved is quite ordinary: anthropologists interpret behaviours – verbal behaviours in particular – by
attributing beliefs, desires, and intentions to individual or collective
actors, in a manner that makes these behaviours appear rational.

One might assume that the best interpretation is the most faithful
one – that is, the one whose content most resembles that of the
interpreted representation. On reflection, however, things are not
that simple. If her aim were just to maximize faithfulness, the
anthropologist would publish only translations of actually uttered
words. However, most utterances heard by the anthropologist make
sense only in the very specific context in which they were produced; they rely on shared cultural representations which they do
not express directly.

The anthropologist must, for her own sake to begin with, go beyond mere translation: only then can she hope to understand what she hears, and thus be genuinely able to translate it. She must speculate, synthesize, reconceptualize. The interpretations that the anthropologist constructs in her own mind or in her notebooks are too complex and detailed to be of interest to her future readers; moreover, they tend to be formulated in an idiosyncratic jargon in which native terms, technical terms used in an *ad hoc* way, and personal metaphors mix freely. Later, writing for readers who will spend a few hours on a study to which she has devoted years, the anthropologist must synthesize her own syntheses, retranslate her own jargon, and, unavoidably, depart even more from the details conveyed by her hosts. In order to be more relevant, the anthropologist must be less faithful.

Moreover, similarity of content varies with the point of view and the context. To say, for instance, that for the Txikao, the human body 'anabolizes strong substances' is suggestive and not misleading in the context of Menget's discussion: in that context, the notion of anabolization is taken quite metaphorically. In other words, the resemblance between the chemical notion of anabolization and the Txikao notion it interprets is seen as pertinent, but quite restricted. On the other hand, the same interpretive statement would be misleading in the context of a comparative study of cultural views of the chemistry of digestion, where considerations of relevance would lead one to take the notion of anabolization much more literally.

The intuitive, context-dependent character of interpretation does not entail that all interpretations are equally good or bad; but it does entail that that our criteria of evaluation are themselves partly intuitive and of limited intersubjective validity. Some imaginable interpretations would be, by all reckonings, quite bad: for example, that the true content of the Holy Trinity dogma is a recipe for chocolate mousse. But it may happen that significantly different interpretations of the same representation all seem plausible. The data interpreted by Menget in an 'intellectualist' manner (i.e. as involved in an attempt at explaining the world) might, for instance, be approached with equal subtlety in a psychoanalytic vein. Presented with both types of interpretations, readers would, no doubt, choose according to their theoretical preferences. Moreover, in doing so, they would act rationally.

But here is the rub: if it is rational to prefer one particular interpretation to another on the basis of prior theoretical preferences, then it is hard – if not impossible – to validate or invalidate a general theory on the basis of particular interpretations.

Interpretation allows a form of understanding that we cannot do without in everyday life: the understanding of representations, mental and public, and therefore the understanding of people. In the scientific study of representations, interpretation is just as indispensable a tool as it is in everyday life. But can we use as a scientific tool an intuitive, partly subjective form of understanding?

No evidence is absolutely reliable, and, arguably, no evidence is theory-independent. However, the basic requirement for the scientific use of any evidence is not that it should be absolutely reliable and theory-independent, but only that it should be more reliable than the theories that it serves to confirm or disconfirm and therefore independent of those particular theories (or of any equally controversial theory).

Some interpretations are more reliable than others, and more intersubjectively acceptable. If these interpretations somehow depend on 'theories' of human comprehension, these are tacit theories that human beings in general, and anthropologists in particular, are not even aware of, and therefore are not intending to challenge. Thus, we would all, I presume, be disposed to trust Menget and accept his claim that Opote could not consume his catch without the risk of damaging his fishing magic as, at the very least, a reasonably approximate interpretation of part of what Opote himself or others around him might have said. That is, we would trust Menget's ability to understand and sometimes to anticipate what individual Txikao may have said to him on specific occasions, just as we would trust ourselves if we had been in Menget's place, having learnt the language, spent the time among the Txikao, and so forth. Fairly literal, flat interpretations of particular utterances and ordinary intentions made by interpreters competent in the language and familiar with the people are not totally reliable or theory-independent, but they are often uncontroversial.

Commonsensical interpretations of particular utterances and of other normally intelligible individual behaviours are reliable enough to be used, with methodological caution, as basic evidence for anthropological theorizing. That is, these interpretations are significantly more reliable than the theories we might want to test with their help. On the other hand, more speculative forms of interpretation, such as interpretations of beliefs that believers themselves are incapable of articulating, or interpretations of 'collective mentalities', whatever their attractions and merits, will not do as evidence.

The question then is: Can anthropological theorizing rely only on the first, more reliable, but also more modest, kind of interpretation? The answer depends on the kind of theorizing one wants.

Explaining Cultural Representations

'To explain' may be taken in two senses. In the first one, to explain a cultural representation – for instance, a sacred text – is to make it intelligible – that is, to interpret it. The previous section dealt with such interpretive explanations. In the second sense, to explain a cultural representation is to show how it results from relatively general mechanisms at work in a given specific situation. In this second sense, the only one to be considered in this section, the explanation of cultural representations has an essential theoretical aspect: the identification of the general mechanisms at work. This theoretical objective is not a concern of most anthropologists, whose main focus is ethnography, and is pursued only in a scattered, piecemeal fashion. There is not even a majority view – let alone any general agreement – as to what might be regarded as an adequate explanatory hypothesis in anthropology.

Simplifying greatly (and with apologies for the unfairness that such simplification necessarily entails), I will nevertheless distinguish four types of explanation – or purported explanation – in anthropology, three of them widespread: interpretive generalization, structuralist explanation, and functionalist explanation; and a rarer type of explanation, a version of which I have been defending for some time: epidemiological models.

Interpretive generalizations

Many anthropologists seem to think that *a* – if not *the* – right way to arrive at theoretical hypotheses consists in taking the interpretation of some particular phenomenon in a given culture and progressively generalizing it to all phenomena of the same type in all cultures, by taking into account increasingly diverse data.

The very idea of 'couvade', for instance, is the outcome of an interpretive synthesis of quite disparate behaviours. Various 'theories' of the couvade differ, on the one hand, with respect to the manner in which they synthesize the evidence, and on the other hand, with respect to the other phenomena which they see as crucially related to the couvade. Thus, on the basis of European examples, the couvade was long considered as a symbolic – more precisely, hyperbolic – way for the father to claim for himself some of the advantages of motherhood. Mary Douglas, for instance, suggests: 'The couvading husband is saying, "Look at me, having cramps and contractions even more than she! Doesn't this prove I am the father of her child?" It is a primitive proof of paternity' (1975: 65).

Claude Lévi-Strauss offers another generalized interpretation, inspired by native American examples:

> It would be a mistake to suppose that a man is taking the place of the woman in labour. The husband and wife sometimes have to take the same precautions because they are identified with the child who is subject to great dangers during the first weeks or months of its life. Sometimes, frequently for instance in South America, the husband has to take even greater precautions than his wife because, according to native theories of conception and gestation, it is particularly his person which is identified with that of the child. In neither event does the father play the part of the mother. He plays the part of the child. (Lévi-Strauss 1966a: 195)

Patrick Menget, whose essay develops Lévi-Strauss's suggestion, concludes in a more abstract fashion (rendered even more abstract by out-of-context quotation): 'The power of the couvade lies in its articulation of a logic of the natural qualities of the human being and a problematic of succession, and in signifying by its progression and durability the irreversibility of human time' (Menget 1982: 208).

Such anthropological interpretations raise two issues. First, what exactly are these interpretations supposed to represent? Some would say that they represent the general meaning of the institution they interpret. Yet, any bearer of meaning, be it a text, a gesture, or a rit-ual, does not bear meaning in itself, but only *for someone*. For whom, then, does the institution have its alleged meaning? Surely, it must be for the participating people, say for Opote and his fellows. There is every reason to suppose, however, that the participants take a view of their institution that is richer, more varied, and more linked to local considerations than a transcultural interpretation could ever hope to express. At best, therefore, these general interpretations are a kind of decontextualized condensation of very diverse local ideas: a gain in generality means a loss in faithfulness.

The second issue raised by these interpretive generalizations is the following: in what sense do they *explain* anything? How – and for whom – would the performance of an easy rite by the husband of every about-to-be or new mother serve as a 'proof of paternity'? How would the father's playing 'the part of the child' protect – or even seem to protect – the child from grave dangers? Who would willingly endure great deprivations for the sake of 'signifying . . . the irreversibility of human time'? A meaning is not a cause; and the attribution of a meaning is not a causal explanation. (Of course, there are cases where the attribution of a meaning to a behaviour fills a gap in an otherwise satisfactory causal explanation, but not so here.)

Interpretive generalizations don't explain anything, and aren't, properly speaking, theoretical hypotheses. Interpretive generaliza-tions are patterns that can be selected, rejected and modified at will in order to construct interpretations of local phenomena. As such, and only as such, may they be useful.

Structuralist explanations

Structuralist explanations attempt to show that the extreme diversity of cultural representations results either from variations on a small number of underlying themes, or from various combinations of a finite repertory of elements, or from regular transformations of underlying simple structures.

All varieties of structural analysis start from interpretive general-izations, but then attempt to go beyond them. This rooting of struc-tural analysis in interpretive generalization is particularly manifest in the work of one of the founders of the genre, Georges Dumézil (e.g. 1968). Dumézil tried to show that the myths and rituals of Indo-Europeans are all variations on the same underlying pattern: an image of social life as constituted of three 'functions': sovereignty, war, and production. This tri-functional pattern is, of course, an interpretive generalization; but Dumézil exploited it in a properly structuralist way. He tried to show how this pattern gave rise to different structural developments according to the type of cultural phenomena involved (pantheons, myths, epics, rituals, etc.) and according to the particular culture. He did not search for the explanation of this common pattern and varying structural developments in interpretation, but rather in history, building on the model of historical linguistics.

In Dumézil's style of structural analysis, just as in standard interpretive generalizations, the only relationships among representations held to be relevant are relationships of resemblance: two representations resembling one another can both be interpreted by means of a third representation which abstracts from their differences. Lévi-Strauss (e.g. 1963, 1973) widened the field of structural analysis by considering that systematic differences are no less relevant than resemblances.[4] He maintained, for instance, that a myth may derive from another myth not just by imitating it, but also by systematically reversing some of its features: if, say, the hero of the first myth is a giant, the hero of the second myth might be a dwarf; if the one is a killer, the other might be a healer, and so on. Thus a network of correspondences richer than mere resemblance relationships may be discovered among representations: either among representations of the same type – myths, for instance – or between different types of representations – myths and rituals, for instance.

Menget takes a Lévi-Straussian line when he attempts to relate the couvade to the prohibition of incest. The couvade, as he interprets it, expresses the progressive separation of the child's substance from that of its parents. Incest prohibition prevents a man and a woman descended from the same parents from re-fusing substances separated by means of the couvade:

> There is both a relationship of continuity between the couvade and the incest prohibition, since the latter keeps separated what the former had separated out of a common substance, and a functional

complementarity, insofar as the couvade orders a communication within the social group which allow its diversification, and the incest prohibition establishes its external communication. (Menget 1982: 208)

Such a structural account does not by itself explain the couvade; but, if one accepts it, it modifies the explanatory task. The *explanandum* is no longer just the couvade; it is a complex of representations and practices having to do with the mechanism of biological reproduction (as understood by the Txikao), a complex the coherence of which the anthropologist has attempted to establish, in spite of its superficial motley appearance.

Structural analysis raises two main problems, one methodological, the other theoretical. The methodological problem is as follows: in order to establish structural relationships among representations, the anthropologist interprets them. It is among the resulting interpretations, rather than among the observable or recordable data, that systematic resemblances and differences may be manifest. However, with a bit of interpretive ingeniousness, any two complex objects can be put into such a structural relationship. One could thus show that, say, Hamlet and Little Red Riding Hood are in a relationship of 'structural inversion':

Hamlet	*Little Red Riding Hood*
A male hero	A female hero
hostile to his mother	obeying her mother
meets a terrifying, supra-human creature (the Spectre)	meets a reassuring infra-human creature (the wolf)
who is in fact well-disposed	who is in fact ill-disposed
and who tells him not to waste time	and who tells her to take her time
etc.	etc.

Such pastiches do not, of course, invalidate structural analysis. But they do illustrate its limits: the reliability of the analysis cannot be greater than that of the interpretations it employs. And the fact is that structuralists, like all other anthropologists, practice interpretation essentially guided by their intuitions and without any explicit methodology. Moreover, the interpreter's intuitions are themselves guided by the aims of structural analysis, with an obvious risk of circularity and no obvious safeguards.

The theoretical problem raised by structural analysis is the following: in what way does structural analysis constitute an explanation of cultural phenomena? Some defenders of structuralism see in their approach a mere means of putting order into the data – that is, a means of classifying, rather than explaining. Dumézil combined structural analysis with historical explanation. Lévi-Strauss combines in a more intricate manner structural analysis with an essentially psychological kind of genetic explanation. The structures uncovered through structural analysis are assumed to be the product of a human mind inclined to flesh out abstract structures with concrete experience and to explore possible variations on these structures.

For instance, a given cultural group makes use of representations of animal characters in order to display in the form of a myth some basic conceptual contrasts: say, between nature and culture, descent and affinity, life and death. A neighbouring group may then transform the myth, by reversing the value of some of the characters, thus symbolizing, over and above the contents of the myth, the group's difference from the neighbours from whom the myth was actually borrowed. Progressive transformations of the myth from one group to another may render it unrecognizable; but the systematic character of these transformations makes it possible for structural analysis to bring to the fore the underlying common structures, which, ultimately, are supposed to be the structures of the human mind. Lévi-Strauss himself has hardly tied his investigations to those of contemporary psychology. The mental mechanisms deemed to generate cultural representations are postulated, but not described.

More generally, the theoretical problem raised by structural analysis boils down to this: complex objects, such as cultural phenomena, display all kinds of properties. Most of these properties are epiphenomenal: they result from the fundamental properties of the phenomenon, but are not among those fundamental properties. In particular, they play no causal role in the emergence and development of the phenomenon, and are not, therefore, explanatory. Structural analysis brings to the fore some systematic properties of phenomena, but, in itself, provides no means of distinguishing fundamental properties from epiphenomenal ones. In a nutshell, structural analysis does not explain; at best, it helps to clarify what needs to be explained.

Functionalist explanations

Showing that a cultural phenomenon has beneficial effects for the social group has been a favourite form of 'explanation' in anthropology. Functional analyses differ according to the type of beneficial effects (biological, psychological, or sociological) they stress. In the Marxist improved version of functional analysis (see Bloch 1983 for a review), contrary effects and dysfunctions are taken into account in order to throw light on the dynamics of society.

Functional analyses have been a great source of sociological insight. However, they are all subject to two objections, one well known and having to do with their explanatory power, the other less well known and having to do with their use of interpretations.

Might a description of the effects of a cultural phenomenon provide an explanation of this phenomenon? Yes in principle, but with two qualifications: first, the effects of a phenomenon can never explain its emergence; second, in order to show how the effects of the phenomenon explain its development, or at least its persistence, one must establish the existence of some feedback mechanism.

Let us suppose that a given cultural institution – for instance, the couvade – has beneficial effects for the groups that have adopted it. For this to help explain the presence of some form of couvade in so many cultures, it needs to be shown that these beneficial effects significantly increase the chances of survival of the cultural groups that are, so to speak, 'carriers' of the institution, and, thereby, the chances of the institution persisting. The onus of proof is, of course, on the defenders of such a functional explanation.

In practice, most functionalists are content to show, often with great ingeniousness, that the institutions they study have some beneficial effects. The existence of an explanatory feedback mechanism is hardly ever discussed, let alone established. Imagine, for instance, a functionalist, taking as her starting-point an interpretation of the couvade similar to that proposed by Mary Douglas. She could argue easily enough that the couvade strengthens family ties – in particular, those between a father and his children, and thereby enhances social cohesion. But how would she go from there to an explanatory feedback mechanism? Moreover, it would not be too hard to establish that many institutions, including the couvade, have harmful effects: food deprivation, such as that suffered by Opote and his fellows, may in some cases be quite damaging.

Most cultural institutions do not affect the chances of survival of the groups involved to an extent that would explain their persistence. In other words, for most institutions a description of their functional powers is not explanatory. Even when such a description does provide some explanatory insight, it does so in a very limited manner: the feedback mechanism explains neither the introduction of cultural forms through borrowing or innovation, nor the transformation of existing cultural forms.

Another, less noticed weakness of the functionalist approach is that it fails to provide any specific principle for the identification of types of cultural phenomena. Rather, it relies uncritically on an interpretive approach.[5]

What is it, for instance, that is supposed to make different local practices tokens of the same general type, say the 'couvade', a type which the anthropologist must then try to describe and explain? The identification of types is never based on function alone: for instance, no one would argue that all the sundry practices that have the 'function' of strengthening father-children ties should be seen as constituting a distinct, homogeneous anthropological type. The identification of types is not based on behaviour: some behaviours may count as couvade in one society and as individual neurosis in another. In fact, whatever its function, whatever its behavioural features, a practice is categorized as an instance of couvade only in accordance with the native point of view. However, native points of views are local, and quite diverse even within the same culture. So, in the end, the identification of a cultural type is based on the synthetic anthropological interpretation of a motley of local interpretations.

Thus 'couvade' is defined by means of an interpretive generalization: local practices that *can be interpreted as* ritual precautions to be taken by a prospective or new father are classified as cases of couvade. As I argued before, the price for such an interpretive usage is a heavy loss of faithfulness: the conception of a ritual, that of an appropriate precaution, what it means for a practice to be imposed on someone, who is considered a father, and so on vary from culture to culture. At the level of generality adopted by anthropologists in their 'theoretical' work, these local conceptions can be interpreted in indefinitely many ways. A few interpretations are retained by anthropological tradition; most local variations and other interpretive possibilities are simply ignored.

Is the loss of faithfulness with respect to local representations compensated for by a gain in relevance? More specifically, are the types defined by means of such interpretive generalizations useful types for theoretical work? I see no reason to believe that they are. Why should one expect all tokens of an interpretively defined type to fall under a common, specific functional explanation – or, for that matter, under any common, specific causal explanation? The point is not particular to the couvade; it holds for all cases of interpretively defined institutions – that is, for all the types of institutions defined in anthropology. From a causal-explanatory point of view, anthropological typologies, being based on interpretive considerations, are quite arbitrary.

Epidemiological models

We call 'cultural', I suggested, those representations that are widely and durably distributed in a social group. If so, then there is no boundary, no threshold, between cultural representations and individual ones. Representations are more or less widely and durably distributed, and hence more or less cultural. Under such conditions, to explain the cultural character of some representations amounts to answering the following question: Why are these representations more successful than others in a given human population? And in order to answer this question, the distribution of all representations must be considered.

The causal explanation of cultural facts amounts, therefore, to a kind of *epidemiology of representations*. An epidemiology of representations will attempt to explain cultural macro-phenomena as the cumulative effect of two types of micro-mechanisms: individual mechanisms that bring about the formation and transformation of mental representations, and inter-individual mechanisms that, through alterations of the environment, bring about the transmission of representations.

From an epidemiological perspective, what the anthropologist refers to as the Txikao couvade amounts to a recurring causal chain of individual thoughts and behaviours. To explain the phenomenon, so understood, would consist in identifying the psychological and ecological factors sustaining this causal chain. I am unable to offer an epidemiological explanation of the Txikao couvade, and I doubt that the ethnographic data, which were collected in a quite different theoretical framework, would provide the evidence required to corroborate any such explanation. I would like to indicate however, from an epidemiological point of view, what kind of questions are raised by such a case, and what kind of answers one would have to look for.

Associated to foreseeable risks, such as perinatal mortality, one often finds rituals aimed at protecting against these risks. From an epidemiological point of view, the recurrence of dangers of a given type is an ecological factor capable of stabilizing a ritual practice. Any practice conceived as a defence against a recurrent type of danger is regularly re-actualized by the danger itself. However, this is so only if the people concerned have faith in the efficacy of the practice. One cannot explain practices such as the Txikao couvade without explaining the fact that they are seen as effective.

To a large extent, if the Txikao believe in the efficacy of the couvade, it is because, generation after generation, they are born into a world in which this efficacy is taken for granted. In other words, such a belief rests on confidence in the authority of elders. From a cognitive point of view, however, it would be surprising if the observation of cases of misfortune had no effect whatsoever on the strength of the belief. Assuming that the practice is, in fact, quite inefficacious, one should expect the belief in its efficacy to be progressively eroded from one generation to the next, especially given the fact that the practice has blatant disadvantages.

There are four types of cases the consideration of which should, in principle, tell people whether their practice is efficacious or not. They are the cases where:

1a The practice has been strictly followed, and the misfortune did not occur.
1b The practice has been strictly followed, and the misfortune did occur.
2a The practice has not been strictly followed, and the misfortune did not occur.
2b The practice has not been strictly followed, and the misfortune did occur.

Regarding a practice such as the Txikao couvade, which is without real efficacy, the consideration of these four types of cases should, in the long run, convince people that the type of misfortune they fear occurs no less frequently when the practice is strictly followed than when it is not. At the very least, consideration of actual cases should provide no evidence for the efficacy of the practice. There are, then, two possible hypotheses: either people are indifferent to the available evidence, or the inferences they draw from this evidence are unwarranted. These are both cognitive hypotheses, each testable independently of the ethnographic phenomena they might help explain.

It can be shown that not only humans, but also other animals, are capable of spontaneously evaluating specific probabilities and taking them into account, for instance in their foraging behaviour. However, it is also well established that in many situations probabilities are ill-understood, and tend to be distorted in systematic ways.[6]

In general, there are three reasons to expect that an excessive weight may spontaneously be given to cases of type 2b – that is, cases where the failure to adhere to the practice is followed by misfortune. First, only misfortune always begs for an explanation. Second, when failure to adhere to the practice is followed by misfortune, it may appear to have caused it. Third, explaining a misfortune as caused by some people's behaviour makes it possible to assign responsibility, and to give at least a social response to a situation which, otherwise, leaves one powerless. In such conditions, following the practice does protect against at least one risk: that of being accused of having provoked a misfortune. The practice does, quite objectively, have that kind of efficacy.

The cognitive disposition to spontaneously assign excessive weight to cases which have greater relevance in one's life (but not necessarily greater statistical relevance) interacts with an ecological factor: namely, the frequency of different kind of cases. It is likely (and could be ascertained experimentally) that, as the objective frequencies of the four kinds of cases vary with different types of misfortune, it is easier or harder to adequately evaluate the efficacy or inefficacy of the ritual practices involved.

It is harder, for instance, to be mistaken about the efficacy of a practice claimed to protect against a very high risk. One may therefore predict that ineffective practices aimed at preventing unavoidable misfortunes, such as the death of very old people, are subject to a very rapid cognitive erosion. Such practices should be much rarer in human cultures than ineffective practices aimed at preventing misfortunes with an intermediate incidence, such as perinatal mortality in non-medicalized societies. One may also predict that when adherence to an inefficacious practice falls below a specific threshold (which is itself a function of the incidence of the type of misfortune involved), its inefficacy becomes manifest, and the practice either disappears or is radically transformed.

The preceding remarks should help explain why practices aimed at protecting against various types of misfortune can stabilize, even though they are without efficacy. But how is one to explain the specific content of these practices? Why do the Txikao attempt to protect their new-borns by means of food abstinence, rather than by means of songs, or, for that matter, banquets? One might here take as a starting-point interpretive or structuralist studies such as that of Menget. Such analyses tend to show that the Txikao couvade is part and parcel of a coherent ensemble of cultural representations. Even leaving aside the exaggeration of coherence typical of interpretive anthropology, it is true that elements of a single culture tend to cohere to a remarkable degree. In itself, such coherence is something to be explained, rather than an explanation of anything.

Here is what an epidemiological and cognitive perspective suggests. In the process of transmission, representations are transformed. This occurs not in a random fashion, but in the direction of contents that require lesser mental effort and provide greater cognitive effects. This tendency to optimize the effect–effort ratio – and therefore the *relevance* of the representations transmitted (see Sperber and Wilson 1986/1995) – drives the progressive transformation of representations within a given society towards contents that are relevant in the context of one another. The particular content of a practice such as the Txikao couvade will be all the more stable, the more relevant it is in the context of other Txikao cultural representations. To explain the Txikao couvade, therefore, one would have to study the the particular context in which Txikao cognitive and communicative activities take place, trying to identify the factors which, through these activities, stabilize the institution. Thus, the epidemiological approach can interact with standard ethnography, by drawing on answers already provided and by raising novel interrogations.

The epidemiological approach renders manageable the methodological problem raised by the fact that our access to the content of representations is unavoidably interpretive. The solution to this methodological problem of ethnography is not to devise some special hermeneutics giving us access to representations belonging to a culture, yet uninstantiated in the individual heads or the physical environment of its members. The solution is merely to render more reliable our ordinary ability to understand what people like you, Opote or me say and think. This is so because, in an epidemiological explanation, the explanatory mechanisms are individual mental mechanisms and inter-individual mechanisms of communication; the representations to be taken into account are those which are constructed and transformed by these micro-mechanisms. In other words, the relevant representations are at the same concrete level as those that daily social intercourse causes us to interpret.

Another methodological advantage of the epidemiological approach is that it provides a principled way to identify the types of cultural things for which a more general explanation is to be sought. The proper objects for anthropological theorizing are types of causal chains of the kind I have described. These types of causal chains are to be individuated in terms of features that play a causal role in their emergence and maintenance. These features may be ecological or psychological: for instance, the lability of oral texts, as opposed to the stability of written ones, is a key ecological factor in explaining their respective distributions; the high memorability of narratives, as opposed to the low memorability of descriptions, is a key psychological factor. These two factors interact in an obvious way, and justify considering oral narratives as a proper anthropological type.

The psychological features pertinent to determining types of cultural things may well include features of their content. Of course, content features can be characterized only interpretively. To say that various representations share a content feature amounts to saying that they can all be interpreted, at a given level and from a given point of view, in the same way. Still, that property of common interpretability, with all its vagueness, may suffice, if not to describe, then at least to pick out, a class of phenomena all affected by some identical causal factors.

There is, for instance, one type of data that anthropologists systematically collect in the field: genealogies. According to society and social class, the depth of genealogies varies. Some people remember long lines of ancestors, while others can hardly go beyond the generation of their grandparents. As students in anthropology learn in their first year, relationships recognized as genealogical differ from society to society, and are not equivalent to relationships of simple biological descent. Under such conditions, the very notion of a genealogy, as a type of cultural representation, is interpretively defined, and hence, quite vague. Still, it is quite plausible that genealogies, in all their versions, are locally relevant, and hence culturally successful, in part for universal reasons.

In an epidemiological perspective, I suggest, the explanation of a cultural fact – that is, of a distribution of representations – is to be sought not in some global macro-mechanism, but in the combined effect of countless micro-mechanisms. What are the factors that lead an individual to express a mental representation in the form of a public representation? What mental representations are the addressees of the public representation likely to construct? What transformation of content is this process likely to bring about? What factors and what conditions render probable the repeated communication of some representations? What properties, either general or contextual, does a representation need in order to maintain a relatively stable content in spite of such repeated communications?

These and other questions raised by an epidemiological approach are difficult; but, at least, anthropologists share many of them with cognitive psychologists, and a relationship of mutual relevance between the two disciplines may emerge and help. In order to answer these questions, as in the case of all anthropological questions, interpretations must be used as evidence. The interpretations required in this approach, however, are of the kind we use all the time in our daily interactions. This does not make them unproblematic, but we should recognize their value as evidence; in fact, we already recognize the evidential value of such interpretations in personal matters much dearer to us than mere scientific theorizing.[7]

3

Anthropology and Psychology: Towards an Epidemiology of Representations

When Malinowski was a student, anthropology and psychology were each well-integrated domains of research, and anthropologists and psychologists could have a command of their whole field. Indeed, many of them – Rivers, Wundt and Malinowski himself, for instance – had a command of both fields. Three-quarters of a century later, the situation is quite different: anthropology and psychology are no longer domains of research, but families of such domains, institutional associations of loosely related enterprises. To put it bluntly, 'anthropology' and 'psychology' are less the names of two sciences than of two kinds of university departments.

Anthropologists and psychologists occasionally show interest in each other's work, argue or co-operate. I do not propose to review these sundry interactions; others have done it much better than I could.[8] What I would like to consider here is the relationship between a central concern of anthropology, the causal explanation of cultural facts and a central concern of psychology, the study of conceptual thought processes. In spite of their centrality, neither the explanation of cultural facts, nor the psychology of thought is a well-developed domain. They are, rather, at a programmatic or, at best, exploratory stage. So, perforce, must be a discussion of their relationship.

This chapter originated as the Malinowski Memorial Lecture for 1984, and was first published in 1985 in *Man*, 20, 73–89. Reprinted by permission of the Royal Anthropological Institute of Great Britain and Ireland.

Malinowski maintained that cultural facts are partly to be explained in psychological terms. This view has often met with scepticism, or even scorn, as if it were an easily exposed naïve fallacy. But what I find fallacious are the arguments usually levelled against this view, and what I find naïve is the belief that human mental abilities make culture possible and yet do not in any way determine its content and organization.

The question is not whether psychological explanations of cultural facts are, in principle, admissible. The question is which psychological considerations are, in effect, explanatory. In this respect, the view I would like to defend contrasts with that of Malinowski. He laid emphasis on the psychology of emotions; I, on the psychology of cognition.[9] He saw some cultural representations as based on psychological dispositions and as answering psychological needs (just as he saw other aspects of culture as answering biological needs). I believe that, more important than needs, and at least as important as dispositions, is a psychological *susceptibility to* culture.

Epidemiology

The human mind is susceptible to cultural representations in the same way as the human organism is susceptible to diseases. Of course, diseases are, by definition, harmful, whereas cultural representations are not. Do you believe, though, that every cultural representation is useful, functional, or adaptive? I do not. Some representations are useful, some are harmful; most, I guess, have no outstanding beneficial or detrimental effects on the welfare of the individual, the group, or the species – not the kind of effects which would provide us with an explanation.

What is it, anyway, that we want to explain? Consider a human group. That group hosts a much larger population of representations. Some of these representations are entertained by only one individual for but a few seconds. Other representations inhabit the whole group over several generations. Between these two extremes, one finds representations with narrower or wider distributions. Widely distributed, long-lasting representations are what we are primarily referring to when we talk of culture. There exists, however, no threshold, no boundary with cultural representations on one side and individual ones on the other. Representations are more or less widely and lastingly distributed, and hence more or less cultural. So, to explain culture is to answer the following question: why are some representations more successful in a human population, more 'catching' than others? In order to answer this question, the distribution of representations in general has to be considered.

I see, then, the causal explanation of cultural facts as necessarily embedded in a kind of *epidemiology of representations*.[10] There are, to begin with, some obvious superficial similarities. For instance, a representation can be cultural in different ways: some are slowly transmitted over generations; they are what we call traditions, and are comparable to endemics; other representations, typical of modern cultures, spread rapidly throughout a whole population but have a short life-span; they are what we call fashions, and are comparable to epidemics.

Epidemiologists have constructed sophisticated mathematical models of the transmission of diseases, and it is tempting to try and apply them to various forms of cultural transmission. This is the line taken by Cavalli-Sforza and Feldman (1981). While their work is worth paying attention to, especially given the dearth of explanatory models in the study of culture, they underestimate important differences between the transmission of diseases and cultural transmission. At the same time they fail to appreciate deeper similarities between the epidemiology of diseases and that of representations.

The transmission of infectious diseases is characterized by processes of replication of viruses or bacteria. Only occasionally do you get a mutation instead of a replication. Standard epidemiological models picture the transmission of stable diseases or of diseases with limited and foreseeable variations. Representations, on the other hand, tend to be transformed each time they are transmitted. For instance, your understanding of what I am saying is not a reproduction in your mind of my thoughts, but the construction of thoughts of your own which are more or less closely related to mine. The replication, or reproduction, of a representation, if it ever occurs, is an exception. So an epidemiology of representations is first and foremost a study of their transformations; it considers the reproduction of representations as a limiting case of transformation.

Epidemiology of diseases occasionally has to explain why a disease is transformed in the process of transmission. Epidemiology of representations, by contrast, has to explain why some representations remain relatively stable – that is, why some representations become properly cultural. As a result, if and when we need mathematical models of cultural transmission, I doubt that we can borrow or easily adapt standard epidemiological models. Similar comments would apply to other biological models of culture, such as those put forward by Dawkins (1976) and by Lumsden and Wilson (1981).

It is possible, though, to pursue the epidemiological analogy in a different, more relevant direction. Epidemiology is not an independent science which studies an autonomous level of reality. Epidemiology studies the distribution of diseases, and diseases are characterized by pathology. The distribution of diseases cannot be explained without taking into account the manner in which they affect the organism – that is, without looking at individual pathology and, more generally, at individual biology. Conversely, epidemiology is a major source of evidence and suggestion for pathology.

What pathology is to epidemiology of diseases, psychology of thought should be to epidemiology of representations: I expect the epidemiology of representations, and therefore the causal explanation of cultural facts, on the one hand, and the psychology of thought, on the other, to stand in a relationship of partial interpenetration and mutual relevance.

Most discussions of the relationship between anthropology and psychology, at the theoretical level we are presently considering, have been in terms of reductionism versus anti-reductionism, as if these were truly available alternatives, and the only available alternatives at that. For reductionists, cultural facts are psychological facts to be explained in psychological terms; for anti-reductionists, cultural facts belong to an autonomous level of reality, and have to be explained essentially in terms of one another. I believe that neither reductionism nor anti-reductionism make much sense in this case, and that the epidemiological analogy provides a more plausible approach.

The notion of the reduction of one *theory* to another is fairly well understood; it is illustrated by famous cases such as the reduction of thermodynamics to statistical mechanics (see Nagel 1961: ch. 11).

The notion of the reduction of one field of inquiry to another, such as the reduction of anthropology to psychology, is much vaguer, particularly so when neither field is characterized by a well-established theory. In such cases, assertions to the effect that one field can or cannot be reduced to another are generally based on a priori convictions rather than specific arguments; some people believe in the unity of science, others in emergent evolution. Relationships between fields are, however, too varied and subtle to be analysed solely, or even primarily, in terms of reduction or non-reduction.[11]

Epidemiology, for instance, is the ecological study of pathological phenomena. It is as eclectic in its ontology as ecology is. It has no more ontological autonomy than ecology has. It does not reduce to pathology, yet it cannot be defined or developed independently of pathology. Of course, one could have an epidemiology of good health or of any other condition; or, as I am suggesting, one can have an epidemiology of representations. But whatever 'epidemiology' one is considering, it has to be defined in relationship to some sister discipline.

What I want to suggest with the epidemiological analogy is that psychology is necessary but not sufficient for the characterization and explanation of cultural phenomena. Cultural phenomena are ecological patterns of psychological phenomena. They do not pertain to an autonomous level of reality, as anti-reductionists would have it; nor do they merely belong to psychology, as reductionists would have it.

The epidemiological analogy is appropriate in yet another way. The distribution of different diseases – say malaria, lung cancer and thalassaemia – follows different patterns, and falls under quite different explanations. So, while there is a general epidemiological approach characterized by specific questions, procedures and tools, there is no such thing as a general theory of epidemiology. Each type of disease calls for an *ad hoc* theory, and though analogies are frequent and suggestive, there is no principled limitation on how much different cases might differ. Similarly, the project of a general theory of culture seems to me misguided. Different cultural phenomena – say funerary rituals, myths, pottery and colour classifications – might well fall under quite different explanatory models.

What the epidemiological analogy suggests is a general approach, types of questions to ask, ways of constructing concepts, and a plurality of not too grand theoretical aims.

Representations

The notion of representation is often used in studies of culture, but ever since Durkheim's 'collective representations', it has been left in a kind of ontological haze. This will not do if we seriously want to develop an epidemiology of representations. A representation involves a relationship between three terms: an object is a representation *of* something, *for* some information-processing device. Here, we shall only consider representations *for* human individuals, ignoring other information-processing devices, such as telephones and computers, even though they affect the distribution of representations in human populations. We shall consider representations *of* anything we please: of the environment, of fictions, of actions, representations of representations and so on, ignoring the difficult philosophical problems this raises.

The issue we cannot ignore is this: what kinds of objects are we talking about when we speak of representations? We can talk of representations as concrete, physical objects located in time and space. At this concrete level, we must distinguish two kinds of representations: there are representations internal to the information-processing device – *mental representations*; and there are representations external to the device and which the device can process as inputs – that is, *public representations*.

Take, for instance, the Mornay sauce recipe in a cookery book. It is a public representation, in this case an ink pattern on a piece of paper which can be read – that is, processed as an input of a certain kind. The reader will form a mental representation of the recipe, which he can then remember, forget, or transform, and which he can also follow – that is, convert into bodily behaviour. Or take a mother telling her daughter the story of 'Little Red Riding Hood'. We have here a public representation, in this case a sound pattern which causes the child to construct a mental representation, which she may remember, forget, transform and tell in her turn – that is, convert into bodily, and more specifically vocal, behaviour. At this concrete level, there are millions of tokens of the Mornay sauce recipe, millions of tokens of 'Little Red Riding Hood' – millions, that is, of both public and mental representations.

An epidemiology of representations is a study of the causal chains in which these mental and public representations are involved: the construction or retrieval of mental representations may cause individuals to modify their physical environment – for instance, to produce a public representation. These modifications of the environment may cause other individuals to construct mental representations of their own; these new representations may be stored and later retrieved, and, in turn, cause the individuals who hold them to modify the environment, and so on.

There are, then, two classes of processes relevant to an epidemiology of representations: intra-individual processes of thought and memory and inter-individual processes whereby the representations of one subject affect those of other subjects through modifications of their common physical environment. Intra-individual processes are purely psychological. Inter-individual processes have to do with the input and output of the brain – that is, with the interface between the brain and its environment; they are partly psychological, partly ecological.

Representations can also be considered at a purely abstract level, without referring either to their mental form in human brains or to their public form in perceptible physical patterns. At this abstract level, formal properties of representations can be discussed: we may notice that the Mornay sauce recipe contains that of béchamel sauce, and discuss it as an example of French bourgeois cuisine – another abstraction. We can analyse the plot of 'Little Red Riding Hood', compare it with that of other tales, and try to argue, in a Lévi-Straussian fashion, that 'Little Red Riding Hood' stands in a relationship of symmetrical inversion to, say, 'Jack and the Beanstalk' (more realistically than Hamlet – see chapter 2).

As abstract objects, representations have formal properties, and enter into formal relations among themselves. On the other hand, abstract objects do not directly enter into causal relations. What caused your indigestion was not the Mornay sauce recipe in the abstract, but your host having read a public representation, having formed a mental representation, and having followed it with greater or lesser success. What caused the child's enjoyable fear was not the story of 'Little Red Riding Hood' in the abstract, but her understanding of her mother's words. More to the present point, what caused the Mornay sauce recipe or the story of 'Little Red Riding Hood' to become cultural representations is not — or rather, is not directly — their formal properties; it is the construction of millions of mental representations causally linked by millions of public representations.

There is a relationship between these concrete processes and the formal properties of the representations processed. Formal properties of representations can be considered in two way (which are not incompatible): as properties of abstract objects considered in themselves (a Platonist approach) or as properties that a processing device, a human mind in particular, could attribute to them and exploit (a psychological approach). In other words, formal properties of representations (or at least some of them) can be considered as potential psychological properties. Potential psychological properties are relevant to an epidemiology of representations. One can ask, for instance, what formal properties make 'Little Red Riding Hood' more easily comprehended and remembered — and therefore more likely to become cultural — than, say, a short account of what happened today on the Stock Exchange.

The Platonist approach may be of great intrinsic interest,[12] but it is not the appropriate approach if one is interested in providing a causal explanation of cultural facts. Both mental and public representations have to be considered, and formal properties have to be seen in psychological terms.

Misconceptions

Most discussions of cultural representations, whether in anthropology, in studies of religion, or in the history of ideas, consider them in abstract terms: a myth, a religious doctrine, a ritual instruction, a legal rule or even a technique is discussed without any consideration of the psychological processes it may undergo or of the interplay of its mental and public representations.

Even self-proclaimed materialists discuss representations without consideration of their material existence as psychological stimuli, processes and states. The difference between self-proclaimed materialists and those whom they accuse of idealism is that 'materialists' see representations more as *effects* of material conditions, while 'idealists' see them more as *causes* of material conditions. Both 'materialists' and 'idealists' talk of representations considered in the abstract as entering into causal relationships with the material world; whatever the order of causes and effects favoured, this presupposes a very unsound form of idealistic ontology.

It is conceivable, of course, that causal explanations of cultural facts could be formulated at a fairly abstract level, ignoring thereby the micro-mechanisms of cognition and communication. This is certainly what anthropologists and sociologists have tried to do, linking, for instance, economic infrastructure and religion. But however good it might be, any such explanation would be incomplete. For economic infrastructure to affect religion, it must affect human minds. There are only two ways, one cognitive, the other non-cognitive, in which human minds can be affected. The mind can be affected by stimuli: that is, by very specific modifications of the brain's physical environment. Or the mind can be affected through non-cognitive physical and, in particular, chemical modification of the brain resulting from, say, nutritional deficiency or electro-shock. To show that economic conditions affect religion, one must be capable of showing how economic conditions affect the interaction of brains and environments in either a cognitive or a non-cognitive way. Moreover, this action must be shown to cause the cognitive and behavioural modifications which, at a more abstract level, are described as religion.

For the time being, we have neither convincing general explanations of cultural facts at an abstract level, nor an epidemiology of representations. The question is, therefore, how should we allocate our efforts? Of course, it is a good thing that we each follow our different hunches, and that we do not all give the same answer to that question. In arguing for an epidemiology of representations, I am not wanting to turn all anthropologists into epidemiologists; I merely want to raise an interest in this alternative approach.

Imagine that a successful explanation of cultural phenomena is possible at an abstract level. It would at best be incomplete. It could not replace an epidemiology of representations solidly rooted in psychology, which would have to be developed anyhow. Imagine, now, a successful epidemiology of representations. Of course, for all we know, it might provide only incomplete or needlessly cumbersome explanations of cultural facts. But there is also the possibility that it would encompass all the explanations we need. An epidemiology of representations is certainly necessary, and possibly sufficient, for the causal explanation of cultural facts. I see that as a strong reason for developing the epidemiological approach.

I do not hope, by that argument, to convince anthropologists and sociologists who are quite content to remain at an abstract level and ignore psychological issues. Their attitude is, I would guess, based less on a misconceived ontology than on a misconceived psychology. They might grant that culture has to be implemented psychologically, yet maintain that the human mind is such that this implementation is easily achieved and does not affect the contents of culture.

In most of the literature, intra- and inter-individual processes are assumed, either implicitly or explicitly, to ensure, on the whole, the simple and easy circulation of just any conceivable representation. The possibility that human cognitive and communicative abilities might work better on some representations than on others is generally ignored. The transformations caused by storage and recall are rarely taken into account: it is as if recall were a mere reversal of the effects of storage. Similarly, inter-individual processes are taken to consist in straightforward imitation, or in the automatic encoding and decoding of representations. If these assumptions were correct, the causal micro-mechanisms of the transmission of representations would be of marginal relevance only; any representation could flow unaltered through the channels of social communication, with just a smooth oscillation between indefinitely repeated mental and public forms. An epidemiology of representations would then be dealing with trivial matters. Nothing irretrievable would be lost in considering cultural representations in purely abstract terms. However, spelling out these psychological assumptions is enough to show that they are utterly naïve.

Without even turning to scholarly psychology, each of us knows through personal experience that some representations, say Gödel's proof, are very hard to comprehend, however much we might like to understand them. Some representations, say a number of twenty digits, though not hard to comprehend, are hard to remember. Some deeply personal representations are hard or even impossible to convey without loss and distortion. On the other hand, there are some representations, say the story of 'Little Red Riding Hood' or a popular tune, which we cannot help remembering, even though we might wish to forget them.

What is it that makes some representations harder to internalize, remember, or externalize than others? One might be tempted to answer, 'Their complexity', and to understand 'complexity' as an abstract property of representations. But this answer will not do. A number of twenty digits is not more complex than the story of 'Little Red Riding Hood'; any standard computer can process the former much more easily and with much less memory space than the latter. In fact, while it is easy enough to provide a computer with the text of a version of 'Little Red Riding Hood', it is not clear how we could provide it with the *story* itself. Human beings, on the other hand, remember a story much more easily than a text. So, what is complex for a human brain differs from what is complex for a computer; complexity is not an explanation, but something to be explained. What makes some representations harder to internalize, remember, or externalize than others, what makes them, therefore, more complex for *humans*, is the organization of human cognitive and communicative abilities.

Dispositions and Susceptibilities

In order to suggest how, in an epidemiological perspective, anthropology and psychology can be mutually relevant, I shall introduce a distinction between dispositions and susceptibilities, and, very briefly, go over a few standard issues in the study of culture.

Human, genetically determined cognitive abilities are the outcome of a process of natural selection. We are entitled to assume that they are adaptive: that is, that they helped the species survive and spread. This is not to say that all their effects are adaptive.

Some of the effects of our genetic endowment can be described as dispositions, others as susceptibilities, although the distinction is not always easy to draw. Dispositions have been positively selected in the process of biological evolution; susceptibilities are side-effects of dispositions. Susceptibilities which have strong adverse effects on adaptation get eliminated with the susceptible organisms. Susceptibilities which have strong positive effects may, over time, be positively selected and become, therefore, indistinguishable from dispositions. Most susceptibilities, though, have only marginal effects on adaptation; they owe their existence to the selective pressure that has weighed, not on them, but on the disposition of which they are a side-effect. Both dispositions and susceptibilities need appropriate environmental conditions for their ontogenetic development. Dispositions find the appropriate conditions in the environment in which they were phylogenetically developed. Susceptibilities may well reveal themselves only as a result of a change in environmental conditions.

Homo sapiens, for instance, has a disposition to eat sweet food. In the natural environment in which the species developed, this was of obvious adaptive value in helping individuals to select the most appropriate nutrients. In the modern environment, in which sugar is artificially produced, this brings out a susceptibility to over-consumption of sugar, with all its well-known detrimental effects.

Basic Concepts

With the distinction between dispositions and susceptibilities in mind, let us consider, first, the problems raised by systems of concepts. Each culture is characterized by a different system of concepts. It is an anthropological problem how much systems of concepts can vary from culture to culture. Are there, that is, universal constraints on the structure of these systems? It is a psychological problem how concepts are formed in individual minds.

One view of concept formation, which has inspired componential analysis in anthropology,[13] and early studies of concept formation in psychology,[14] is that a new concept is formed by combining several previously available concepts. For instance, if a child already has the concept of a parent and that of a female, she can form the concept of a mother by combining 'female' and 'parent'.

On this view of concept formation, concepts which cannot be decomposed into more elementary ones cannot have been acquired, and must therefore be innate. Now, most of our concepts cannot be so decomposed: try to decompose, for instance, 'yellow', 'giraffe', 'gold', 'electricity', 'lackadaisical', 'dignity'. You cannot? Then, on this theory, these concepts and hundreds or thousands more, must be innate, which, except for a few such as 'yellow', seems wildly implausible. Moreover, even when a concept can be formed by combining more elementary ones, there may be reasons to doubt that this is the way its formation actually takes place: surely, children do not form the concept of a mother by constructing the intersection of 'female' and 'parent'. Rather, they form the concept of a parent by constructing the union of 'mother' and 'father'.

Another way in which concepts might be taught and learned is by ostension. You show a child a bird; you tell her, 'This is a bird'; and after a few such experiences, she acquires the concept of a bird. Ostension raises well-known problems: you may well point in the direction of a bird, but you are simultaneously pointing in the direction of a material object, an animal, a crow, this particular crow, a feathered body, the underside of a bird, a thing on a tree, a source of noise, a black thing and an infinity of other things. How is the child to realise that what you intend to draw her attention to is only one of these things, and that the word you utter corresponds to only one of these concepts?

Logical combination and ostension are not mutually incompatible, though. Some admixture of them might provide a more plausible hypothesis. Ostension works if it operates under strong logical constraints. Imagine that a child, without having an innate concept of a bird, has an innate schema for zoological concepts and an innate disposition to apply and develop this schema whenever she is provided with information which seems relevant to the task. Thus, if you point to an animal and utter a word, then unless the context suggests otherwise, the child's first hypothesis will be that you are providing her with a name corresponding to a zoological concept, and more specifically to a zoological taxon. She will expect the concept she is to develop to have the logical properties characteristic of taxonomic concepts. If you were behaving according to her expectations, then she will be on the right track (and if you were not, what kind of a parent are you?).

The anthropological or epidemiological implications of this view of concept formation are clear: humans have a disposition to develop concepts such as that of a bird; as a result, such concepts are 'catching'. It takes remarkably little experience and prompting for children to develop them and apply them appropriately; and once they are present in a language, they are not easily lost. A wealth of such concepts is therefore found in every language.

Let me speculate more generally. I assume that we have an innate disposition to develop concepts according to certain schemas. We have different schemas for different domains: our concepts of living kinds tend to be taxonomic; our concepts of artefacts tend to be characterized in terms of function; our concepts of colour tend to be centred on focal hues; and so on. Concepts which conform to these schemas are easily internalized and remembered. Let us call them *basic concepts*. A large body of basic concepts is found in every language. Of course, basic concepts differ from one language to another, but they do not differ very much. The basic concepts of another language tend to be comparatively easy to grasp, learn and translate.

There is a growing body of research on basic concepts both in psychology and in anthropology, with more collaboration between the two disciplines in this domain than in any other.[15] This work tends to show that individual concept formation, and therefore cultural variability, are indeed governed by innate schemas and dispositions.

This has been shown, of course, only for a few semantic domains. Could it be generalized? Are all concepts formed according to fairly specific innate schemas? I doubt it very much. First, there is no a priori reason to assume that concept formation is always achieved in the same way and therefore falls under a single model. Second, while some concepts are easily acquired with very little prompting, which suggests that there is a readiness for their acquisition, the formation of other concepts, say scientific or religious ones, takes a great deal of time, interaction and even formal teaching. These elaborate concepts are acquired within the framework of complex representations of the world. These representations and, therefore, the concepts which are characteristic of them, are based as much or more on susceptibilities than on dispositions.

Cultural Representations

The social development and individual formation of representations of the world are the next issues I would like to comment upon from an epidemiological point of view. Human cognitive abilities act, among other things, as a filter on the representations capable or likely to be widely distributed in a human population – that is, capable or likely to become cultural representations. In a way, this filtering function has long been recognized. It is generally accepted among anthropologists that an adequate account of a culture's beliefs must show them to be somehow rational in their context.

What is meant by rationality is neither clear nor constant. As generally understood, however, rationality implies a certain degree of consistency between beliefs and experience and among beliefs. Rationality, then, presupposes cognitive mechanisms which tend to prevent or to eliminate empirical inconsistencies and logical contradictions.

Many anthropologists, from Durkheim to Clifford Geertz, have explicitly or implicitly assumed that all the beliefs of a culture, whether banal or mysterious, are mentally represented in the same mode, and therefore achieve rationality in the same way. In our terms, they are filtered by the same cognitive mechanisms. Holders of this view, when they want to explain apparently irrational beliefs, tend to turn to cognitive relativism: the hypothesis that criteria of rationality vary from culture to culture.

Other anthropologists[16] have insisted that everyday empirical knowledge of the world – say, the representation that honey is sweet – and religious beliefs – such as the dogma of the Holy Trinity – and scientific hypotheses – such as the theory of relativity – are not the same kind of mental objects. Different types of representation achieve rationality in different ways. They are cognitively filtered by different processes.

Let me briefly contrast everyday empirical knowledge with religious beliefs. l assume that we have a disposition to develop a certain form of empirical knowledge which could be characterized as follows:

- It consists in representations which are simply stored in encyclopaedic memory and which are treated by the mind as true descriptions of the world just because they are so stored.
- These representations are formulated in the vocabulary of basic concepts; thus, you cannot have this kind of knowledge about atoms, viruses, mana or democracy (which, I assume, do not fall under basic concepts).
- They are automatically tested for mutual consistency and, in particular, for consistency with perceptual inputs.

Everyday empirical knowledge is developed under strong constraints: conceptual, logical and perceptual. As a result, such knowledge tends to be empirically adequate and consistent. On the other hand, it applies only to some cognitive domains, and does so rather rigidly.

Other forms of mental representations are developed with greater flexibility and weaker filtering mechanisms. They involve other cognitive abilities, in particular that of forming representations of representations.

Humans can mentally represent not just environmental and somatic facts, but also some of their own mental states, representations and processes. The human internal representation system – the language of thought, to use Jerry Fodor's expression (Fodor 1975) – can serve as its own metalanguage.

This meta-representational ability, as we might call it, is essential to human acquisition of knowledge (and also to verbal communication, although I will not discuss this here). First, it allows humans to doubt and to disbelieve. Doubting and disbelieving involve representing a representation as being improbable or false. Presumably, other animals do not have the ability to disbelieve what they perceive or what they decode.

Secondly, meta-representational abilities allow humans to process information which they do not fully understand, information for which they are not able at the time to provide a well-formed representation. If an information-processing device without meta-representational abilities finds itself unable to represent some information by means of a well-formed formula of its internal language, then it cannot use or retain the information at all. A device with meta-representational abilities, on the other hand, can embed a defective representation in a well-formed meta-representation.

Children use this ability all the time to process half-understood information. They are told things that they do not quite understand by speakers whom they trust. So they have grounds to believe that what they are told is true, even though they do not know exactly what it is that they are told. A child is told, for instance, that Mr So-and-so has died, but she does not yet have any concept of death. The best representation she can form is defective, since it contains a half-understood concept. In order to process that defective representation, she has to meta-represent it – that is, embed it in a representation of the form, 'it is a fact that Mr So-and-so has "died", whatever "died" means'.

This allows the child to retain the information, even though she does not fully understand it. It also gives her an incentive to develop the concept of death, and, at the same time, provides her with a piece of relevant evidence for the development of this concept. Adults too, of course, when meeting new concepts and ideas that they only half-understand, embed them in meta-representations .

Humans have, I assume, a disposition to use their meta-representational abilities to expand their knowledge and their conceptual repertoire. Meta-representational abilities, however, also create remarkable susceptibilities. The obvious function served by the ability to entertain half-understood concepts and ideas is to provide intermediate steps towards their full understanding. But it also creates the possibility of conceptual mysteries, which no amount of processing could ever clarify, invading human minds.

Rational constraints on half-understood ideas are not very stringent: the internal consistency of a half-understood idea and its consistency with other beliefs and assumptions cannot be properly tested: if any inconsistency appears, it may be due to a mistaken interpretation of the belief. To the child, the very idea of death and, therefore, the claim that someone is dead may seem self-contradictory; yet she may nevertheless, and without irrationality, accept them on the assumption that the fault is with her understanding rather than with the concept or the claim. With half-understood ideas, what is known as the 'argument of authority' carries full weight.

The fact that mysterious ideas and concepts can easily meet criteria of rationality is not sufficient to guarantee their cultural success. There are infinitely many mysteries competing for mental space, and hence for cultural space. What advantage do the winning mysteries possess? They are, I want to suggest, more evocative and, as a result, more memorable.

Evocation can be seen as a form of problem solving: the problem is to provide a more precise interpretation for some half-understood idea. This is done by searching encyclopaedic memory for assumptions and beliefs in the context of which the half-understood idea makes sense. Sometimes the problem raised by a half-understood idea – for instance, by a crossword clue – is easily solved with a short evocation. In other cases the idea is so poorly understood, and so unrelated to the subject's other mental representations, that there is nowhere for the evocation to start. The most evocative representations are those which, on the one hand, are closely related to the subject's other mental representations, and, on the other hand, can never be given a final interpretation. It is these *relevant mysteries,* as they could be described, which are culturally successful.

Apparently irrational cultural beliefs are quite remarkable: they do not appear irrational by departing slightly from common sense, or by timidly going beyond what the evidence allows. They appear, rather, like downright provocations against common-sense rationality. They include beliefs about creatures which can be in two places at the same time or which can be here, yet remain invisible, thus flatly contradicting universal assumptions about physical phenomena; about creatures which can transform from one animal species to another, thus flatly contradicting universal assumptions about biological phenomena; about creatures which know what happened and what will happen without having to be there or to be told, thus flatly contradicting universal assumptions about psychological phenomena.

Some of these paradoxical beliefs could be given well-formed representations, but then they would have to be rejected on grounds of inconsistency. Moreover, rejecting them would entail another kind of inconsistency: it would be inconsistent with the assumption that the source of the beliefs is trustworthy. Overall consistency can be achieved only by treating these beliefs as mysteries. For, as mysteries, they achieve relevance because of their paradoxical character

– that is, because of the rich background of everyday empirical knowledge from which they systematically depart. By achieving relevance, they occupy people's attention, and become better distributed than representations which are mysterious merely by being obscure.

Attempts to explain religious beliefs and other cultural mysteries in terms of some universal psychological disposition have been unconvincing. I believe they were misguided. Unlike everyday empirical knowledge, religious beliefs develop not because of a disposition, but because of a susceptibility.

Memory and Oral Literature

Up to now, I have considered the role only of cognitive processes of formation of concepts and representations. Other cognitive processes, processes of storage and recall in particular, and processes of communication are no less essential to the explanation of cultural facts. Consider the case of a non-literate society, without schools or other learning institutions. There, most learning is spontaneous. Most mental representations are constructed, and stored and retrieved, without deliberation. I would like to put forward a law of the epidemiology of representations which applies to such a society:

> In an oral tradition, all cultural representations are easily remembered ones; hard-to-remember representations are forgotten, or transformed into more easily remembered ones, before reaching a cultural level of distribution.

This law has immediate application, for instance, to the study of oral narratives. We can take it for granted that tales, myths and so on are optimal objects for human memory, or else they would have been forgotten. What is it about these narratives that makes them so memorable? What is it about human memory that makes it so good at remembering these tales? Here the mutual relevance of psychology and anthropology should be obvious. Yet the anthropological study of oral literature is, with a few exceptions,[17] done without concern for psychology. In cognitive psychology, on the other hand, there is a growing body of research on the structure of narratives and its effect on memory,[18] but little or no advantage is taken of anthropological expertise.

When new communication technologies appear, writing in particular, more things can be communicated, and internal memory is supplemented by external memory stores.[19] As a result, memorization and communication have weaker filtering effects. For instance, other forms of literature can develop, and the particular forms found in oral tradition need not be maintained at all.

Concluding Remarks

I would like to insist again that I am not offering an epidemiology of representations as a substitute for other anthropological enterprises, but as a further undertaking which I see as essential to the causal explanation of cultural facts and to fruitful relationships between anthropology and psychology. Even so, it might be objected that the scope I am claiming for an epidemiology of representations is too large. It might be pointed out that all the examples I have discussed so far – concepts, beliefs, narratives – concern representations which can be internalized individually, and which are cultural as a result of a great many individuals internalizing them. But what about institutions? Surely, a school, a ritual, a judicial system are cultural things; yet they are not the kind of things that can be internalized by the individual. Do they not fall, then, outside the scope of an epidemiology of representations, and is not the claim that the causal explanation of cultural facts has to be encompassed in such an epidemiology grossly exaggerated?

Well, here is the counter-objection. An epidemiology of representations; does not study representations, it studies distributions of representations (and therefore all the modifications of the environment which are causally involved in these distributions). Cultural classifications, beliefs, myths, and so forth are indeed characterized by homogeneous distributions: closely similar versions of the same representation are distributed throughout a human population. Other cultural distributions are differential: the distribution of some representations in certain ways causes other representations to be distributed in other ways. This, I submit, is characteristic of institutions.

Some sets of representations include representations of the way in which the set should be distributed.

> An institution is the distribution of a set of representations which is governed by representations belonging to the set itself.

This is what makes institutions self-perpetuating. Hence, to study institutions is to study a particular type of distribution of representations. This study falls squarely within the scope of an epidemiology of representations.

Let me end by illustrating this characterization of an institution with an example. Consider the Malinowski Memorial Lecture. It is, as everyone will agree, an institution. A representation was put on paper when the Lecture was first instituted; unwritten additions were made in the course of time. This representation calls for the yearly distribution of invitations, to a speaker on the one hand, to an audience on the other; it represents the speaker distributing to the audience the complex representation called a lecture; it represents the lecturer including in his lecture some deferential references to Malinowski; it represents the lecturer ending the oral representation after an hour or so, so that the, by then, thirsty audience can go for a drink. It represents the lecturer, a few weeks later, submitting a written version of his oral representation, to the journal *Man,* thus ensuring a wider, more lasting distribution of it. When all these representations have been distributed according to one of them, then you have – or, rather, you have had – a Malinowski Memorial Lecture.

4

The Epidemiology of Beliefs

I would like to bring together two sets of speculations: anthropological speculations on cultural representations and psychological speculations on the cognitive organization of beliefs, and to put forward, on the basis of these speculations, fragments of a possible answer to the question: how do beliefs become cultural? I will not apologize for the speculative character of the attempt. At this stage, either the question is answered in a vague, fragmentary and tentative way, or it must be left alone: there is not enough sound theorizing and well-regimented evidence in the domain to do otherwise.

Anthropological Speculations

I use 'cultural representation' in a wide sense: anything that is both cultural and a representation will do. Thus, cultural representations can be descriptive ('Witches ride on broomsticks') or normative ('With fish, drink white wine'); simple, as in the above examples, or complex, like the common law or Marxist ideology taken as a whole; verbal, as in the case of a myth, or non-verbal, as in the case of a mask, or multi-media, as in the case of a Mass.

To begin with, two remarks about the notion of a representation.

This chapter is a revised version of 'The Epidemiology of Beliefs', published in Colin Fraser and George Gaskell (eds), *The Social Psychological Sudy of Widespread Beliefs* (Oxford: Clarendon Press, 1990), 25–44. Reprinted by permission of Oxford University Press.

First, as we saw in chapter 2, 'to represent' is not a two-place predicate: something represents something; it is a three-place predicate:
something represents something for someone. Second, we should
distinguish two kinds of representations: internal, or *mental representations* – for example, memories, which are patterns in the brain and
which represent something for the owner of that brain – and external, or *public representations* – for example, utterances, which are
material phenomena in the environment of people and which represent something for people who perceive and interpret them.[20]

Which are more basic: public or mental representations? Most
cognitive psychologists (see Fodor 1975) see mental representations
as more basic: for public representations to be representations at all,
they must be mentally represented by their users; for instance, an
utterance represents something only for someone who perceives,
decodes and comprehends it – that is, associates with it a (multi-
level) mental representation. On the other hand, mental representations can exist without public counterparts; for instance, many of
our memories (and all or nearly all the memories of an elephant) are
never communicated. Therefore, it is argued, mental representations
are more basic than public ones.

Most social scientists (and also philosophers such as Ludwig
Wittgenstein (1953) and Tyler Burge (1979)) do not agree: they see
public representations as more basic than mental ones. Public representations are observable, both by their users and by scientists,
whereas mental representations, if they exist at all, can only be sur-
mised. More importantly, it is claimed (e.g. by Vygotsky (1965))
that mental representations result from the internalization of public
representations and of underlying systems (e.g. languages and ideologies) without which no representation is possible. But if so, public
representations must be more basic than mental ones. (This denies
mental representations to non-social animals, but holders of this
view don't mind.)

There is an obvious sense in which public representations do
come before mental ones: a child is born into a world full of public
representations, and is bombarded with them from the first moments
of her life. She does not discover the world unaided, and then make
public her privately developed representations of it; rather, a great
many of her representations of the world are acquired vicariously,

not through experience, but through communication, or through a combination of experience and communication. Moreover, her very ability to communicate effectively is contingent upon her acquiring the language and the other communication tools of her community. However, those who see mental representations as basic are not (or should not be) denying this point. What they are (or should be) denying is that public representations could be of use to the child if she did not have, to begin with, some system of mental representations with which to approach the public ones.

Conversely, those who see public representations as basic are not (or should not be) merely making the trivial point that each individual is born into a world full of public representations and crucially relies on them. They are (or should be) claiming not only that the physical shape of public representations is public, outside people's heads, there for people to perceive, but also that the *meaning* of public representations is public, out there for people to grasp. On this view, meaning – the regular relationship between that which represents and that which is being represented – is social before being individually grasped; hence, in the relevant sense, public representations are more basic than mental ones. This leads anthropologists, in particular, to consider that 'culture is public because meaning is' (Geertz 1973:12). Most anthropologists study culture as a system of public representations endowed with public meanings, without any reference to the corresponding mental representations.

I have a bias: I am a materialist – not in the sense this word too often has in the social sciences, where a materialist is one who believes that the economic 'infrastructure' determines the ideological 'superstructure', but in the sense of philosophy and the natural sciences: that all causes and all effects are material. I then wonder: what kind of material objects or properties could public meanings possibly be? I am not persuaded by Geertz when he dismisses the issue thus:

> The thing to ask about a burlesqued wink or a mock sheep raid [two of his examples of public representations], is not what their ontological status is. It is the same as that of rocks on the one hand and dreams on the other – they are things of this world. The thing to ask is what their import is: what it is . . . that, in their occurrence and

through their agency, is getting said. (Geertz 1973:10)

I am not persuaded, because the task of ontology is not so much to say which things are 'of this world' and which are not, as to say in what manner, or manners, things can be of this world; and regarding cultural things, the problem stands.

We understand reasonably well how material things fit into the world; but we don't know that there are immaterial things, and if there are, we don't know how they fit. Hence, for any class of entities – rocks, memories, or cultural representations – a materialist account, if it is available at all, is preferable, on grounds of intelligibility and parsimony.

In the case of mental things such as memories or reasonings, cognitive psychologists accept at least a form of minimal materialism, which has precise methodological implications. From a cognitive point a view, an appropriate description of a mental phenomenon must, *inter alia*, show that this phenomenon can be materially realized. For example, cognitive psychologists may try to show how a reasoning process could be materially implemented on a computer. Thus, with the development of cognitive psychology, we begin to grasp what kind of material objects mental representations might be.

Now, when it comes to cultural representations allegedly endowed with public meanings, whether we pay lip-service to materialism and declare them to be material too or resign ourselves to ontological pluralism, the truth of the matter is that we have no idea in what manner they might be 'things of this world'.

The materialist alternative is to assume that both mental and public representations are strictly material objects, and to take the implications of this assumption seriously. Cognitive systems such as brains construct internal representations of their environment partly on the basis of physical interactions with that environment. Because of these interactions, mental representations are, to some extent, regularly connected to what they represent; as a result, they have semantic properties, or 'meaning', of their own (see Dretske 1981; Fodor 1987b; Millikan 1984). Public representations, on the other hand, are connected to what they represent only through the meaning attributed to them by their producers or their users; they have no semantic properties of their own. In other words, public representations have meaning only through being associated with mental representations.

Public representations are generally attributed similar meanings by their producers and by their users, or else they could never serve the purpose of communication. This similarity of attributed meaning is itself made possible by the fact that people have similar enough linguistic and encyclopaedic knowledge. Similarity across people makes it possible to abstract from individual differences and to describe 'the language' or 'the culture' of a community, 'the meaning' of a public representation, or to talk of, say, 'the belief' that witches ride on broomsticks as a single representation, independently of its public expressions or mental instantiations. What is then described is an abstraction. Such an abstraction may be useful in many ways: it may bring out the common properties of a family of related mental and public representations; it may serve to identify a topic of research. Mistake this abstraction for an object 'of this world', however, and you had better heed Geertz's advice and ignore its ontological status.

From a materialist point of view, then, there are only mental representations, which are born, live and die within individual skulls, and public representations, which are plain material phenomena – sound waves, light patterns and so forth – in the environment of individuals. Take a particular representation – witches on broomsticks – at an abstract level: what it corresponds to at a concrete level is millions of mental representations and millions of public representations the meanings of which (intrinsic meanings in the case of mental representations, attributed meanings in the case of public ones) are similar to that of the statement: 'Witches ride on broomsticks'. These millions of mental and public representations, being material objects, can and do enter into cause–effect relationships. They may therefore play a role both as *explanans* and as *explanandum* in causal explanations. The materialist wager is that no other causal explanation of cultural phenomena is needed.

Consider a human group: it hosts a much wider population of representations. Some of these representations are constructed on the basis of idiosyncratic experiences, as, for instance, my memories of the day on which I stopped smoking; others are based on common experiences, as, for instance, our belief that coal is black; others still derive from communication rather than from direct experience, as, for instance, our belief that Shakespeare wrote *Macbeth*. Common experience and communication bring about a similarity of representations across individuals; or, loosely speaking, they cause some representations to be shared by several individuals, sometimes by a whole human group. This loose talk is acceptable only if it is clear that when we say that a representation is 'shared' by several individuals, what we mean is that these individuals have mental representations similar enough to be considered versions of one another. When this is so, we can produce a further version – a public one this time – to identify synthetically the contents of these individual representations.

When we talk of cultural representations – beliefs in witches, rules for the service of wines, the common law, or Marxist ideology – we refer to representations which are widely shared in a human group. To explain cultural representations, then, is to explain why some representations are widely shared. Since representations are more or less widely shared, there is no neat boundary between cultural and individual representations. An explanation of cultural representations, therefore, should come as part of a general explanation of the distribution of representations among humans – as part, that is, of an *epidemiology of representations*.

The idea of an epidemiological approach to culture is by no means new. It was suggested by Gabriel Tarde (1895, 1898). Several contemporary biologists have developed it in various ways. The value of an epidemiological approach lies in making our understanding of micro-processes of transmission and macro-processes of evolution mutually relevant. However, if the micro-processes are fundamentally misunderstood, as I believe they have been in previous epidemiological approaches, the overall picture is of limited value. Whatever their differences and their merits, past approaches share a crucial defect: they take the basic process of cultural transmission to be one of replication, and consider alterations in transmission as accidents.

The view of cultural transmission as a process of replication is grounded not only in a biological analogy – a genic mutation is an accident, replication is the norm – but also in two dominant biases in the social sciences. First, as we have already seen, individual differences are idealized away, and cultural representations are too often treated as identical across individuals throughout a human group or subgroup. Second, the prevailing view of communication, as a coding process followed by a symmetrical decoding process, implies that replication of the communicator's thoughts in the minds of the audience is the normal outcome of communication.

In *Relevance: Communication and Cognition*, Deirdre Wilson and I have criticized this code model of human communication, and developed an alternative model which gives pride of place to inferential processes. One of the points we make – a commonsensical point, really, which would hardly be worth making if it were not so often forgotten – is that what human communication achieves in general is merely some degree of resemblance between the communicator's and the audience's thoughts. Strict replication, if it exists at all, should be viewed as just a limiting case of maximal resemblance, rather than as the norm of communication. (The same is also true of imitation, another well-known but little understood means of cultural transmission, which I won't go into here.) A process of communication is basically one of transformation. The degree of transformation may vary between two extremes: duplication and total loss of information. Only those representations which are repeatedly communicated *and* minimally transformed in the process will end up belonging to the culture.

The objects of an epidemiology of representations are neither abstract representations nor individual concrete representations, but, we might say, strains, or families, of concrete representations related both by causal relationships and by similarity of content. Some of the questions we want to answer are: what causes such strains to appear, to expand, to split, to merge with one another, to change over time, to die? Just as standard epidemiology does not give a single general explanation for the distribution of all diseases, so there is no reason to expect that these questions will be answered in the same way for every kind of representation. The diffusion of a folktale and that of a military skill, for instance, involve different cognitive abilities, different motivations and different environmental factors. An epidemiological approach, therefore, should not hope for one grand unitary theory. It should, rather, try to provide interesting questions and useful conceptual tools, and to develop the different models needed to explain the existence and fate of the various families of cultural representations.

Though which factors will contribute to the explanation of a particular strain of representations cannot be decided in advance, in every case, some of the factors to be considered will be psychological, and some will be environmental or ecological (taking the environment to begin at the individual organism's nerve endings and to include, for each organism, all the organisms it interacts with). Potentially pertinent psychological, factors include the ease with which a particular representation can be memorized, the existence of background knowledge in relationship to which the representation is relevant, and a motivation to communicate the content of the representation. Ecological factors include the recurrence of situations in which the representation gives rise to, or contributes to, appropriate action, the availability of external memory stores (writing in particular), and the existence of institutions engaged in the transmission of the representation.

Unsurprisingly, psychological and ecological factors are themselves affected by the distribution of representations. Previously internalized cultural representations are a key factor in one's susceptibility to new representations. The human environment is, for a great part, man-made, and made on the basis of cultural representations. As a result, feedback loops are to be expected both within models explaining particular families and between such models. The resulting complexity is of the ecological rather than of the organic kind. Though 'organicism' has disappeared from the anthropological scene, the organicist view of a culture as a well-integrated whole still lingers. The epidemiological approach departs from such cultural holism; it depicts individual cultures as wide open, rather than almost closed, systems and as approximating an ecological equilibrium among strains of representations, rather than as exhibiting an organic kind of integration. It is then of interest to find out which strains of representations benefit one another, and which, on the contrary, compete.

The identification of epidemiological phenomena in classical epidemiology often arises out of the study of individual pathology, but the converse is also true: the identification of particular diseases is often aided by epidemiological considerations. Similarly, when types of mental representations have been identified at a psychological level, the question of their epidemiology arises; and, conversely, when particular strains of representations, or mutually supportive strains, have been epidemiologically identified, the question of their psychological character arises. More generally, as with the pathology and epidemiology of diseases, the psychology and epidemiology of representations should prove mutually relevant.

Psychological Speculations

Anthropologists and psychologists alike tend to assume that humans are rational – not perfectly rational, not rational all the time, but rational enough. What is meant by rationality may vary, or be left vague, but it always implies at least the following idea: humans beliefs are produced by cognitive processes which are on the whole epistemologically sound; that is, humans approximately perceive what there is for them to perceive and approximately infer what their perceptions warrant. Of course, there are perceptual illusions and inferential failures, and the resulting overall representation of the world is not totally consistent; but, as they are, the beliefs of humans allow them to form and pursue goals in a manner which often enough leads to the achievement of those goals.

Anthropologists and psychologists tend to assume that humans are rational, without explaining why. I assume some degree of rationality because it makes good biological sense. Why did vertebrates evolve so as to have more and more complex cognitive systems, culminating, it seems, in the human one, if not because this makes their interaction with the environment (e.g. feeding themselves, protecting themselves) more effective? Now, only an epistemologically sound cognitive system (i.e. one that delivers approximations of knowledge rather than pretty patterns or astounding enigmas) can serve that purpose, and, for that, it must be rational enough. This way of explaining why humans are rational implies that there is an objective reality, and that at least one function of human cognition is to represent in human brains aspects of that reality.

Fitting together reality and reason in this manner may seem commonsensical to psychologists, but many anthropologists – not so long ago, most of them – know better. People of different cultures have beliefs which are not only very different, but even mutually incompatible. Their beliefs from our point of view, ours from theirs, seem irrational. If we want to maintain, nevertheless, that both they and we are rational, then an obvious way out is to deny that there is an objective reality to begin with. Reality on that view is a social construct, and there are at least as many 'realities' or 'worlds' as there are societies. Different beliefs are rational in different socially constructed worlds. I have argued at length against this view (see Sperber 1974, 1985b). Here I will merely state my bias: I find a plurality of worlds even less attractive than a plurality of substances; if there is a way, I would rather do without it.

There is a way, but first we must do a bit of conceptual house cleaning.[21] What are we referring to when we talk of 'beliefs'? Take an example: we tend to assume that Peter believes that it will rain if he says so, or assents to somebody else saying so, or, in some cases, if he takes his umbrella on his way out. We do not mistake these behaviours for the belief itself; we take them, rather, as caused in part by Peter having the belief in question and, therefore, as evidence of the belief. We might be tempted to say, then (as many philosophers have – e.g. Ryle 1949), that a belief is a disposition to express, assent to, or otherwise act in accordance with, some proposition. As psychologists, however, we will want to go deeper and find out what kind of mental states might bring about such a disposition. An answer often heard nowadays is that humans have a kind of 'data base' or 'belief box' (Steven Schiffer's phrase) in which some conceptual representations are stored.[22] All representations stored in that particular box are treated as descriptions of the actual world. When the occasion is right, this yields the usual behavioural evidence for belief: assertion and assent in particular.

The belief box story, however attractive, cannot be the *whole* story. Many of the propositions to which we are disposed to assent are not represented at all in our mind – a well-known point – and many of the propositions we are disposed not only to assent to but also to express and, in some cases, to act in accordance with are not, or not simply, stored in a data base or belief box – a more controversial point.

You have long believed that there are more pink flamingos on Earth than on the Moon, but no mental representation of yours had, until now, described that state of affairs. We may well have an infinity of such unrepresented beliefs, and a large proportion of these are widely shared, though of course they have never been communicated. It is reasonable, however, to assume that what makes them unrepresented beliefs (more specifically, propositions to which we are disposed to assent) is that they are inferable from other beliefs which *are* mentally represented. What we need to add for this to the belief box is some inferential device allowing subjects to accept as theirs these unrepresented beliefs on the basis of the actually represented ones. The inferences in question are not made consciously, so the inferential device hooked up to the belief box must be distinct from, and need not resemble, human conscious reasoning abilities (see Sperber and Wilson 1986: ch. 2).

Besides accounting for unrepresented beliefs, hooking the belief box up to an inferential device introduces a factor of rationality in the construction of beliefs. Suppose that some of the representations in our belief box come from perception (broadly understood to include the 'perception' of one's own mental states), and that all other beliefs are directly or indirectly inferred from the perceptually based ones. This will already ensure areas of consistency among our beliefs. Suppose, furthermore, that the inferential device recognizes an inconsistency when it meets one, and corrects it. Then you get a tendency to enlarge areas of consistency (even though contradictory beliefs may still be held, provided they are never used as joint premisses in an inference).

While perception plus unconscious inference might be the whole story for the beliefs of elephants, it could not be for the beliefs of humans. There are two interconnected reasons for this: first, many – possibly most – human beliefs are grounded not in the perception of the things the beliefs are about, but in communication about these things. Second, humans have a meta-representational, or *interpretive*, ability. That is, they can construct not only *descriptions* – that is, representations of states of affairs – but also *interpretations* – that is, representations of representations.[23] Now, humans use this interpretive ability to understand what is communicated to them and, more generally, to represent meanings, intentions, beliefs, opinions, theories and so on, whether or not they share them. In particular, they can represent a belief and take a favourable attitude to it, and therefore express it, assent to it, and generally show behaviours symptomatic of belief, on a basis quite different from belief box inclusion.

Young Lisa is told by her teacher: 'There are male and female plants.' She understands 'male' and 'female' with respect to animals as more or less an extension of the distinction between men and women: females have children, males fight more easily, and so on. She does not see in plants anything resembling this distinction, and so she does not quite understand what her teacher is telling the class. On the other hand, she understands it in part; she understands that in some species there are two types of plants, and guesses that this difference has to do with reproduction, and so on. She trusts her teacher, and if he says that there are male and female plants, then she is willing to say so herself, to say that she believes it, and to exhibit various behaviours symptomatic of that belief.

Behind Lisa's belief behaviour, do we have a genuine belief? Not of the belief box kind, certainly, since such a half-understood idea (what I called a 'semi-propositional representation' in Sperber 1985b: ch. 2) could not have emerged from perception or from inference from perception: it is a typical outcome, rather, of not totally success-ful communication. Remember, too, that the inferential device must be able to operate freely on beliefs in the belief box so as to yield more mutually consistent beliefs; but in that case half-understood ideas should not be allowed directly in the box, since their consistency with other representations and their implications are largely indeterminate.

But how, then, might Lisa's half-understood idea of there being male and female plants be represented in her mind? Well, she might have in her belief box the following representations:

What the teacher says is true.
The teacher says that there are male and female plants.

Lisa's partial understanding of 'there are male and female plants' is now embedded in a belief box belief about what her teacher said. This belief, together with the other belief that 'what the teacher says is true', provides a validating context for the embedded representa-tion of the teacher's words. This gives Lisa rational grounds for exhibiting many of the behaviours symptomatic of belief – but grounds quite different from plain belief box inclusion.

What this example suggests is that the beliefs we attribute to peo-ple on the evidence provided by their behaviour do not belong to a single psychological kind; in other words, quite different types of mental states can bring about identical belief behaviour.

I maintain that there are two fundamental kinds of beliefs represented in the mind. There are descriptions of states of affairs directly stored in the belief box; let us call this first kind *intuitive beliefs*. Such beliefs are intuitive in the sense that they are typically the product of spontaneous and unconscious perceptual and inferential processes; in order to hold these intuitive beliefs, one need not be aware of the fact that one holds them, and even less of reasons for holding them. Then there are interpretations of representations embedded in the validating context of an intuitive belief, as in the above example; let us call this second kind *reflective beliefs*. These beliefs are reflective in the sense that they are believed in virtue of second-order beliefs about them.[24]

Intuitive beliefs are derived, or derivable, from perception by means of the inferential device. The mental vocabulary of intuitive beliefs is probably limited to *basic concepts*: that is, concepts referring to perceptually identifiable phenomena and innately pre-formed, unanalysed abstract concepts (of, say, norm, cause, substance, species, function, number, or truth). Intuitive beliefs are on the whole concrete and reliable in ordinary circumstances. Together they paint a kind of common-sense picture of the world. Their limits are those of common sense: they are fairly superficial, more descriptive than explanatory, and rather rigidly held.

Unlike intuitive beliefs, reflective beliefs do not form a well-defined category. What they have in common is their mode of occurrence: they come embedded in intuitive beliefs (or, since there can be multiple embeddings, in other reflective beliefs). They cause belief behaviours because, one way or another, the belief in which they are embedded validates them. But they may differ in many ways: a reflective belief may be half-understood but fully understandable, as in the above example of the sex of plants; or, as I will shortly illustrate, it may remain half-understood for ever; or, on the contrary, it may be fully understood. The validating context may be an identification of the source of the reflective belief as a reliable authority (e.g. the teacher) or an explicit reasoning. Given the variety of possible contextual validations for reflective beliefs, commitment to these beliefs can widely vary, from loosely held opinions to fundamental creeds, from mere hunches to carefully thought-out convictions. Reflective beliefs play different roles in human cognition, as I will very briefly illustrate.

For Lisa, forming and storing the half-understood reflective belief that there are male and female plants may be a step towards a more adequate understanding of the male–female distinction. It provides her with an incomplete piece of information which further encounters with relevant evidence may help complete. After she achieves an adequate understanding of the matter, her reflective belief that there are male and female plants may well be transferred to, or duplicated in, her belief box as an intuitive belief. So, one role of reflective belief is to serve as a 'hold' format for information that needs to be completed before it can constitute an intuitive belief.

Now, consider the following case. Young Bobby has in his belief box the two representations:

What Mom says is true.
Mom says that God is everywhere.

Bobby does not fully understand how somebody, be it God or anyone else, can be everywhere. However his mother saying so gives him sufficient ground to exhibit all the behaviours symptomatic of belief: he will readily state that God is everywhere, will assent when the same statement is made by others, and may even refrain from sinning in places where (apparently) nobody can see him. That God is everywhere is for Bobby a reflective belief. As he grows older, he may keep this belief and enrich it in many ways, but, if anything, its exact meaning will become even more mysterious than it was at first. Here is a belief which, like most religious beliefs, does not lend itself to a final, clear interpretation, and which therefore will never become an intuitive belief. Part of the interest of religious beliefs for those who hold them comes precisely from this element of mystery, from the fact that you are never through interpreting them. While the cognitive usefulness of religious and other mysterious beliefs may be limited (but see Sperber 1975b), it is not too difficult to see how their very mysteriousness makes them 'addictive'.

In the two examples considered so far – Lisa and the sex of plants,

Bobby and divine omnipresence – what made the reflective representation a belief was the authority granted to the source of the representation: the teacher and the mother respectively. Laymen accept scientific beliefs on authority too. For instance, most of us believe that $e = mc^2$ with only a very limited understanding of what this formula means, and no understanding of the arguments that led to its adoption. Our belief, then, is a reflective belief of mysterious content, justified by our trust in the community of physicists. It is not very different, in this respect, from Bobby's belief that God is everywhere.

There is a difference, though. Even for theologians, that God is everywhere is a mystery, and they too accept it on authority. For physicists, on the other hand, the theory of relativity is not a mystery, and they have reasons to accept it which have nothing to do with trust. Well-understood reflective beliefs, such as the scientific beliefs of scientists, include an explicit account of rational grounds to hold them. Their mutual consistency and their consistency with intuitive beliefs can be ascertained, and plays an important, though quite complex, role in their acceptance or rejection. Still, even for physicists, the theory of relativity is a reflective belief; it is a theory, a representation kept under scrutiny and open to revision and challenge, rather than a fact that could be perceived or unconsciously inferred from perception.

Half-understood or mysterious reflective beliefs are much more frequent and culturally important than scientific ones. Because they are only half-understood and therefore open to reinterpretation, their consistency or inconsistency with other beliefs, intuitive or reflective, is never self-evident, and does not provide a robust criterion for acceptance or rejection. Their content, because of its indeterminacy, cannot be sufficiently evidenced or argued for to warrant their rational acceptance. But that does not make these beliefs irrational: they are rationally held if there are rational grounds to trust the source of the belief (e.g. the parent, the teacher, or the scientist).

This, then, is my answer to those who see in the great diversity and frequent apparent inconsistency of human beliefs, an argument in favour of cultural relativism: there are two classes of beliefs and they achieve rationality in different ways. Intuitive beliefs owe their rationality to essentially innate, hence universal, perceptual and inferential mechanisms; as a result, they do not vary dramatically, and are essentially mutually consistent or reconcilable across cultures. Those beliefs which vary across cultures to the extent of seeming irrational from another culture's point of view are typically reflective beliefs with a content that is partly mysterious to the believers themselves. Such beliefs are rationally held, not in virtue of their content, but in virtue of their source. That different people should trust different sources of beliefs – I, my educators, you, yours – is exactly what you would expect if they are all rational in the same way and in the same world, and merely located in different parts in this world.

Different Types of Beliefs, Different Mechanisms of Distributions

Let us now bring together the anthropological and psychological speculations developed so far. If there are different kinds of beliefs, then we might expect them to be distributed by different mechanisms. More precisely, we might expect the distribution of intuitive beliefs, which are a relatively homogeneous kind, to proceed along roughly common lines,[25] and the distribution of reflective beliefs, which are much more diverse, to take place in many different ways. In this concluding section, I would like to suggest that such is indeed the case.

In all human societies, traditional or modern, with or without writing, with or without pedagogic institutions, all normal individuals acquire a rich body of intuitive beliefs about themselves and their natural and social environment. These include beliefs about the movement of physical bodies, the behaviour of one's own body, the effects of various body–environment interactions, the behaviour of many living kinds, the behaviour of fellow humans. These beliefs are acquired in the course of ordinary interaction with the environment and with others. They need no conscious learning effort on the part of the learner and no conscious teaching effort on the part of others (see Atran and Sperber 1991). Even without teaching, these beliefs are easily acquired by everybody. The more fundamental ones are acquired quite early, suggesting a very strong innate predisposition (see Keil 1979; Carey 1982, 1985; Gelman and Spelke 1981; Hirschfeld 1984, 1994).

Some intuitive beliefs are about particulars (particular locations, personal events, individual animals or people), and are idiosyncratic or are only shared very locally; others are general (or about widely known particulars such as historical events and characters), and are widespread throughout a society. General intuitive beliefs vary across cultures, but they do not seem to vary greatly. To mention just one piece of anecdotal evidence, one has yet to find a culture in which where intuitive beliefs about space and movement are so different from modern Western ones that the natives have inordinate problems in learning to drive a car. Much recent work in ethnoscience shows, too, that cross-cultural differences in zoological, botanical, or colour classification are rather superficial, and that for each of these domains (and presumably for other domains, e.g. artefacts or mental states), there are underlying universal structures (see Berlin and Kay 1969; Berlin et al. 1973; Berlin 1978; Atran 1985, 1986, 1987).

What role does communication play in the construction of intuitive beliefs? The answer is not simple. Intuitive beliefs are (or are treated as) the output of perception and unconscious inference, either the subject's own perceptions and inferences or those of others in the case of intuitive beliefs acquired through communication. Even when an intuitive belief is derived from the subject's own perceptions, the conceptual resources and the background assumptions which combine with the sensory input to yield the actual belief have, in part, been acquired through communication. So, it seems, both perception and communication are always involved in the construction of intuitive beliefs. Perception is involved either as the direct source of the belief or as its assumed indirect source (which puts a strong constraint on the possible contents of intuitive beliefs). Communication is involved either as a direct source or, at the very least, a source of concepts and background.[26]

What, now, is the relationship between the relative shares of perception and communication in the construction of an intuitive belief, on the one hand, and its social distribution, on the other? Is it the case that the greater the share of communication, the wider the distribution? Again, the answer is not that simple. A great number of very widespread beliefs owe their distribution to the fact that all

members of a society, or in some cases all humans, have similar perceptual experiences. However, as already suggested, the resources
for perception are themselves partly derived from communication.

Take the widespread intuitive belief that coal is black: were you
told it, or did you infer it from your own perception? Hard to
know. But even if you inferred it from perception, in doing so, you
used the concepts of black and of coal, and how did you acquire
those? Regarding 'black', it seems that the category is innately pre-
wired, so that, when you learned the word 'black', you merely
acquired a way to express verbally a concept you already possessed
(see Berlin and Kay 1969; Carey 1982). Regarding 'coal', no one
would claim that the concept is innate; but what might well be
innate is the structure of substance-concepts with the expectation of
regular phenomenal features – in particular, colour. So, while you
probably acquired the *concept* of coal in the process of learning the
word 'coal', acquiring the concept meant no more than picking the
right innate conceptual schema and fleshing it out. In the process of
fleshing it out, either you were told, or you inferred from what you
saw, that coal is black.

It does not make much difference, then, whether an individual's
belief that coal is black is derived from perception or from communication: once the concept of coal is communicated, the belief that
coal is black will follow one way or the other. This is generally true
of widespread intuitive beliefs. These beliefs conform to cognitive
expectations based on culturally enriched innate dispositions, and are
richly evidenced by the environment. As a result, different direct
perceptual experiences and different vicarious experiences acquired
through communication converge on the same general intuitive
beliefs.

Widespread intuitive beliefs, even exotic ones, are rarely surprising. They are not the kind of beliefs that generally excite the curi-
osity of social scientists, with the exception of cognitive
anthropologists. Among psychologists, only developmental psychologists have started studying them in some detail. Yet intuitive
beliefs not only determine much of human behaviour; they also
provide a common background for communication and for the
development of reflective beliefs.

Whereas widespread intuitive beliefs owe their distribution both to common perceptual experiences and to communication, wide-spread reflective beliefs owe theirs almost exclusively to communi-cation. The distribution of reflective beliefs takes place, so to speak, in the open: reflective beliefs are not only consciously held; they are also often deliberately spread. For instance, religious believers, polit-ical ideologists, and scientists, however they may differ otherwise, see it as incumbent upon them to cause others to share their beliefs. Precisely because the distribution of reflective beliefs is a highly visi-ble social process, it should be obvious that different types of reflec-tive beliefs reach a cultural level of distribution in very different ways. To illustrate this, let us consider very briefly three examples: a myth in a non-literate society, the belief that all men are born equal, and Gödel's proof.

A myth is an orally transmitted story which is taken to represent actual events, including 'supernatural' events incompatible with intuitive beliefs. Therefore, for a myth to be accepted without inconsistency, it has to be insulated from intuitive beliefs: that is, held as a reflective belief. A myth is a cultural representation; this means that the story is told (given public versions) often enough to cause a large enough proportion of a human group to know it (have mental versions of it). For this, two conditions must be met. First the story must be easily enough and accurately enough remembered on the basis of oral inputs alone. Some themes and some narrative structures seem in this respect to do much better cross-culturally than others. The changing cultural background affects memorability, too, so that the content of a myth tends to drift over time so as to maintain maximal memorability.

Second, there must be enough incentives to actually recall and tell the story on enough occasions to cause it to be transmitted. These incentives may be institutional (e.g. ritual occasions where telling the story is mandatory); but the surest incentive comes from the attractiveness of the story for the audience and the success the story-teller can therefore expect. Interestingly, though not too sur-prisingly, the very same themes and structures which help one remember a story seem to make it particularly attractive.

If the psychological conditions of memorability and attractiveness are met, the story is likely to be well distributed; but in order for it to be a myth, rather than, say, a mere tale recognized and enjoyed as such, it must be given credence. What rational grounds do people have to accept such a story as true? Their confidence in those who tell it to them: typically, their confidence in elders whom they have many good reasons to trust and who themselves claim no other authority than that derived from *their* elders. The originator of the chain might be a religious innovator who claimed divine authority for a distinctly different version of older myths. Reference to elders provides a self-perpetuating authority structure for a story which already has a self-perpetuating transmission structure. Still, the authority structure is more fragile than the transmission structure, and many myths loose their credibility, though neither their memorability nor their attractiveness, and end up as tales.

The belief that all men are born equal is a typically reflective belief: it is not produced by perception or by unconscious inference from perception. Rather, except for a few philosophers who originated the belief, all those who have held it came to it through communication. Such a belief does not put any significant weight on memory, but it does present a challenge for understanding, and indeed it is understood differently by different people. As already suggested, the fact that it lends itself to several interpretations probably contributed to its cultural success.

Still, the most important factor in the success of the belief that all men are born equal is its extreme relevance – that is, the wealth of its contextual implications (see Sperber and Wilson 1986) – in a society organized around differences in birthrights. People who accepted, and indeed desired, the implications of this belief found there grounds to accept the belief itself and to try to spread it. However, there was a risk, not to holding the belief, but to spreading it, and so the belief spread only where and when there were enough people willing to take this risk. In other words, unlike a myth, which seems to have a life of its own and to survive and spread, as myth or as tale, in a great variety of historical and cultural conditions, the cultural destiny of a political belief is tied to that of institutions. Ecological factors (more particularly, the institutional environment) play a more important role in explaining the distribution of a political belief than cognitive factors.

Consider now a mathematical belief, such as Gödel's proof. Again, all those who hold it, except Gödel himself, arrived at it through communication. However, the communication, and hence the diffusion, of such a belief meets extraordinary cognitive difficulties. Only people with a high enough level of education in mathematical logic can begin to work at understanding it. Outside scholarly institutions, both the means and the motivation to do that work are generally lacking. On the other hand, once the difficulties of communication are overcome, acceptance is no problem at all: to understand Gödel's proof is to believe it.

The human cognitive organization is such that we cannot understand such a belief and not hold it. To some significant extent, and with obvious qualifications, this is the case with all successful theories in the modern natural sciences. Their cognitive robustness compensates, so to speak, for their abstruseness in explaining their cultural success. The fact that successful scientific theories impose themselves on most of those who understand them is manifest to people who don't understand them. This leads, quite rationally, to lay persons believing that these theories are true and expressing as beliefs whatever they can quote or paraphrase from them. Thus Gödel's proof, and scientific theories generally, become cultural beliefs of a different tenor, accepted on different grounds by the scientists themselves and by the community at large.

We might contrast our three examples in the following way. The distribution of a myth is determined strongly by cognitive factors, and weakly by ecological factors; the distribution of political beliefs is determined weakly by cognitive factors, and strongly by ecological factors; and the distribution of scientific beliefs is determined strongly by both cognitive and ecological factors. However, even this exaggerates the similarities between the three cases: the cognitive factors involved in myth and in science and the ecological factors involved in politics and in science, are very different. The very structure of reflective beliefs, the fact that they are attitudes to a representation, rather than directly to a real or assumed state of affairs, allows endless diversity.

Notwithstanding their diversity, explaining cultural beliefs, whether intuitive or reflective, and if reflective, whether half-understood or fully understood, involves looking at two things: how they are cognized by individuals and how they are communicated within a group; or to put it in the form of a slogan: *Culture is the precipitate of cognition and communication in a human population.*

5

Selection and Attraction in Cultural Evolution

Suppose we set ourselves the goal of developing mechanistic and naturalistic causal explanations of cultural phenomena. (I don't believe, by the way, that causal explanations are the only ones worth having; interpretive explanations, which are standard in anthropology, are better at answering some of our questions.) A causal explanation is mechanistic when it analyses a complex causal relationship as an articulation of more elementary causal relationships. It is naturalistic to the extent that there is good ground to assume that these more elementary relationships could themselves be further analysed mechanistically down to some level of description at which their natural character would be wholly unproblematic.

The kind of naturalism I have in mind aims at bridging gaps between the sciences, not at universal reduction. Some important generalizations are likely to be missed when causal relationships are not accounted for in terms of lower-level mechanisms. Other valuable generalizations would be lost if we paid attention to lower-level mechanisms only. If we want bridges, it is so as to be able to move both ways.

Social sciences explanations are sometimes mechanistic, but they are hardly ever naturalistic (with a few exceptions in demography and in historical linguistics). They fail to be naturalistic if only

This was first delivered at a Darwin Seminar at the London School of Economics in May 1995 and as an invited lecture at the International Congress for Methodology and Philosophy of Science in Florence in August 1995. It is to be published in the proceedings of the congress.

because they freely attribute causal powers to entities such as institutions or ideologies the material mode of existence of which is left wholly mysterious. If we want to develop a naturalistic programme in the social sciences, we must exert some ontological restraint and invoke only entities the causal powers of which can be understood in naturalistic terms.

Here is a proposal: let us recognize only human organisms in their material environment (whether natural or artificial), and focus on these organisms' individual mental states and processes and on the physical-environmental causes and effects of these mental things.[27] Here is how, having so restricted our ontology, we might approach the social. A human population is inhabited by a much wider population of mental representations: that is, objects in the mind/brain of individuals such as beliefs, fantasies, desires, fears, intentions and so on. The common physical environment of a population is furnished with the public productions of its members. By 'public production' I mean any perceptible modification of the environment brought about by human behaviour. Productions include bodily movements and the outcomes of such movements. Some productions are long-lasting, like clothes or buildings; other are ephemeral, like a grin or the sounds of speech.

Typically, public productions have mental representations among their causes and among their effects. Mental representations caused by public productions can in turn cause further public productions, that can cause further mental representations, and so forth. There are thus complex causal chains where mental representations and public productions alternate. Public productions are likely to have many mental representations among their causes, and, conversely, every link in a causal chain may be attached to many others, both up and down the causal path.

Of particular interest are causal chains from mental representations to public productions to mental representations and so on, where the causal descendants of a representation resemble it in content. The smallest ordinary such causal chain is an act of successful communication. Typically, the public productions that are involved in communication are *public representations* such as linguistic utterances. Public representations are artefacts the function of which is to ensure a similarity of content between one of their mental causes in the communicator and one of their mental effects in the audience.

Communication is one of the two main mechanisms of transmission, imitation being the other. Transmission is a process that may be intentional or unintentional, co-operative or non-co-operative, and which brings about a similarity of content between a mental representation in one individual and its causal descendant in another individual. Most mental representations are never transmitted. Most transmissions are a one-time local affair. However, it may happen that the recipient of an act of transmission becomes a transmitter in turn, and the next recipient also, and so on, thus producing a long chain of transmission and a strain of mental representations (together with public representations in cases of communication) linked both causally and by similarity of content. Fast-moving rumours and slow-moving traditions are paradigmatic examples of such cultural causal chains.

The Selection Model

When you have a strain of representations similar enough in content to be seen as versions of one another, it is possible and often useful to produce yet another public version in order to represent in a proto-typical manner their partly common content. Thus we talk of *the* belief in metempsychosis, *the* recipe for Yorkshire pudding, *the* story of King Arthur, each identified by a content. These are, of course, abstractions, at least as much so as *the* zebra, *the* Doric order, or *the* Russian peasant. It is tempting to see all the concrete representations that can be identified by means of a prototypical version as having the same content, with only negligible variations, thus as imperfect replicas of one another, but replicas nevertheless. Once this is done, it is but a step to seeing all tokens of the 'same' representation as forming a distinct class of objects in the world, just as all zebras are commonly seen as forming a natural kind. Granting such unity to strains of representations makes it possible to use, in order to develop a causal explanation of culture, one of intellectual history's most powerful tools: the Darwinian idea of selection.

On this approach, cultural representations are self-replicating representations. They replicate by causing those who hold them to produce public behaviours that cause others to hold them too.

Occasionally representations 'mutate', possibly starting a new strain. The task of explaining the contents and evolution of a given culture can be seen, then, as one of finding out which representations are most successful at replicating, under what conditions, and why. Versions of this idea have been defended by, among others, Karl Popper, Donald Campbell, Jacques Monod, Cavalli-Sforza and Feldman, Boyd and Richerson, William Durham, and by Richard Dawkins who coined the name 'memes' for cultural replicators.[28] The success of the word 'meme' is such that it could be seen as confirming, or at least illustrating, the very idea of a meme. I will focus this discussion on Dawkins's memes, espoused in philosophy by Daniel Dennett (1991, 1995) and developed in anthropology by William Durham (1991). My argument extends unproblematically (*mutatis mutandis*, of course) to all the other proposals mentioned.[29]

I have long argued that there is a severe flaw in attempting to develop a naturalistic explanation of cultural evolution on the basis of the Darwinian model of selection. I am not moved by any reservation concerning Darwinism. Quite the opposite, I believe that Darwinian considerations have a central role to play in the explanation of human culture by helping us to answer the fundamental question: what biological and, in particular, what brain mechanisms make humans cultural animals with the kinds of culture they have? In other words, to characterize 'human' in the phrase 'human culture', we must draw on biology, hence on evolutionary theory, hence on the Darwinian model of selection. It is 'culture' in the phrase 'human culture' that calls for some different and, I believe, novel thinking, albeit remaining, as we will see, within the broad Darwinian approach.

My two basic points over the years, and in preceding chapters of this book, have been (1) that representations don't in general replicate in the process of transmission, they transform; and (2) that they transform as a result of a constructive cognitive process. Replication, when it truly occurs, is best seen as a limiting case of zero transformation. These remarks of mine have been taken as an emphatic way of making a correct, but unimportant, point to the effect that replication is not perfect.[30] But after all, hasn't Dawkins himself pointed out that 'no copying process is infallible', and that 'it is no part of the definition of a replicator that its copies must all be perfect' (Dawkins 1982: 85)?

Dawkins, however, is aware of the problem that looms. He writes:

> The copying process is probably much less precise than in the case of genes: there may be a certain 'mutational' element in every copying event. . . . Memes may partially blend with each other in a way that genes do not. New 'mutations' may be 'directed' rather than random with respect to evolutionary trends. . . . These differences may prove sufficient to render the analogy with genetic natural selection worthless or even positively misleading. (Dawkins 1982: 112)

Dawkins's main interest and most relevant contribution are to point out that the mechanism of Darwinian selection is by no means limited to biological material, but may apply to replicators of any substance and any kind.[31] Computer viruses are successful replicators (alas!) of a non-biological kind. Here is an example of an unquestionable cultural replicator. Every now and then, I receive in the mail a chain-letter saying something like:

> Make ten copies of this letter and send them to ten different people. This chain has been started at Santiago de Compostela. Don't break it! Mrs Jones sent ten copies of this letter the very day she received it, and that same week, she won a large prize at the National Lottery. Mr Smith threw away this letter without copying it, and the next day he lost his job.

Here is a text that causes enough of the individuals who receive copies of it to make and send further copies for the process of its distribution to be an enduring one.

Provided that certain conditions obtain, replicators will undergo a process of Darwinian selection. The two main conditions are that there should be variations among replicators, and that different types of replicators should differ in their chances of being replicated. In the case of the selection of genes, the source of variation is random mutation, which is, actually, failure to replicate properly. For selection to operate on replicators capable of mutating, a further condition has to be fulfilled. It has to do with the rate of mutation. Obviously, if genes mutated not just occasionally, but all the time, they wouldn't be replicators any more, and selection would be ineffective. How much mutation is compatible with effective selection? Here is George Williams's answer:

> The essence of the genetical theory of natural selection is a statistical bias in the relative rates of survival of alternatives (genes, individuals, etc.). The effectiveness of such bias in producing adaptation is contingent on the maintenance of certain quantitative relationships among the operative factors. One necessary condition is that the selected entity must have a high degree of permanence and a low rate of endogenous change, relative to the degree of bias. (Williams 1966: 22–3)

In fact, interesting replicators – genes in the biological case – can be characterized as entities replicating well enough to undergo effective selection.[32]

In the case of genes, a typical rate of mutation might be one mutation per million replications. With such low rates of mutation, even a very small selection bias is enough to have, with time, major cumulative effects.[33] If, on the other hand, in the case of culture there may be, as Dawkins acknowledges, 'a certain "mutational" element in every copying event', then the very possibility of cumulative effects of selection is open to question.

There are, of course, bits of culture that do replicate. Some people copy chain-letters. Medieval monks copied manuscripts. Many traditional artefacts are replicas. Thus a pot may be copied by a potter, and some of her pots may be copied by other potters, and so on for many generations of pots and potters. This slow manual reproduction process has been superseded in modern times by more and more sophisticated technologies, such as printing, broadcasting, or e-mail forwarding, which allow massive replication. However, the number of artefactual replicas of a would-be cultural item is only a poor, indirect indicator of its genuine cultural success. Waste-paper baskets and their electronic counterparts are filled with massively replicated but unread junk, while some scientific articles read by only a few specialists have changed our cultural world. The cultural importance of a public production is to be measured not by the number of copies in the environment but by their impact on people's minds.

The most blatant cases of replication are provided by public productions, rather than by mental representations. However, when a public replica is produced by an individual rather than by a machine, this production is caused by an intention or plan of the individual – that is, a mental representation. Mental representations causing the production of public replicas can themselves be seen as mental replicas of mental representations. Jane's mental representation of a pot caused her to make a pot in conformity with this representation. This pot was seen by John, and caused in him the construction of a mental representation identical to Jane's. John's representation caused him to produce a pot identical to Jane's pot, and so on.

The question then arises as to whether the true memes are public productions – pots, texts, songs and so on – that are both effects and causes of mental representations or, as Dawkins (1982) argues, mental representations that are both causes and effects of public productions. With both options, however, there are similar problems. To begin with, most cultural items, be they mental or public, have a large and variable number of mental or public immediate ascendants.

Leaving aside mechanical and electronic reproduction, cases of new items produced by actually copying one given old item are rare. When you sing 'Yankee Doodle', you are not trying to reproduce any one past performance of the song, and the chances are that your mental version of the song was the child of the mental versions of several people. Most potters producing quantities of near-identical pots are not actually copying any one pot in particular, and their skill is typically derived from more than a single teacher (although there may be one teacher more important than the others, which complicates matters further still).

In   general, if you are serious in describing bits of culture – individual texts, pots, songs or individual abilities to produce them – as replications of earlier bits, then you should be willing to ask about any given token cultural item: of which previous token is it a direct replica? In most cases, however, you will be forced to conclude that each token is a replica not of one parent token, nor (as in sexual reproduction) of two parent tokens, nor of any fixed number of parent tokens, but of an indefinite number of tokens some of which have played a much greater 'parental' role than others.

You might want, then, to envisage that this process of synthetic replication of a variable number of models is carried out by a natural equivalent of a morphing programme (i.e. a programme that takes, say, the image of a cat and that of a man as input, and produces the image of a creature somewhere between the cat and the man as output). Just as in a morphing programme, different inputs can be given different weights: you can have your cat-man more like a cat or more like a man, and Jill's skill and her pots may be more like Joan's than like Jane's, though still owing to both Joan's and Jane's skills and pots.

The model that comes to mind now is less immediately reminiscent of the Darwinian notion of selection than of the notion of 'influence' much used in the history of ideas and in social psychology. In the case of selection, genes either succeed or fail to replicate, and sexual organisms either succeed or fail to contribute half the genes of a new organism. Thus relationships of descent strictly determine genic similarity (ignoring mutations). Influence, by contrast, is a matter of degree. Two pottery teachers may have shared the same pupils, and therefore have the same cultural descendants, but their common cultural descendants may be much more influenced by one teacher than by the other. The resulting pots, too, may be descendants of both teachers' pots, but more like the pots of the one than those of the other.

There are, nevertheless, commonalities between the meme model and the influence model. Both involve an idea of competition. Both define a measure of success, in terms of the number of descendants in the one case, in terms of the degree and spread of influence in the other case. Both predict that the most successful items will dominate the culture, and that the culture will evolve as a result of differences in success among competing items. The meme model might be seen as a limiting case of the influence model: the case where influence is either 100 per cent or 0 per cent – that is, where descendants are replicas. Actually, formal models of influence in social psychology tend to concentrate on this limiting case (e.g. Nowak et al. 1990).

Both the meme model and the influence model see human organisms as agents of replication or synthesis, with little or no individual contribution to the process of which they are the locus. At most, the replicating agent may, to some extent, choose what to replicate, and the synthesizing agent may choose not only what inputs to synthesize, but also the weights to give to the different inputs. Among the factors in either reproductive or influential success, then, is the attraction that various possible inputs hold for the agents. However, once inputs (and weights in the case of synthesis) have been chosen, the outcome of a successful process of replication or synthesis is wholly determined. Moreover, on these two views, mental representations involved in cultural transmission never contain more information than the inputs they are supposed to represent or synthesize.

The Attraction Model

The influence model is right, as against the meme model, in treating replication in cultural transmission not as the norm but as a limiting case (of 100 per cent influence). Both are wrong, however, in assuming that, in general, the output of a process of transmission is wholly determined by the inputs (and weights, in the case of influence) accepted or chosen by the receiving organism. The stimulus-drivenness of both models is not the norm of cultural transmission; it too is a limiting case. Not much of culture is transmitted by means of simple imitation or averaging. Medieval monks copying manuscripts – apparently perfect examples of cultural replication – understood what they copied, and, on occasion, corrected what they took to be a mistake in earlier copying on the basis of what they understood. In general, human brains use all the information they are presented with not to copy or synthesize it, but as more or less relevant evidence with which to construct representations of their own.

Let me give three brief illustrations. First, consider your views on President Clinton. They are likely to be very similar to the views of many, and to have been influenced by the views of some. However, it is unlikely that you formed your own views simply by copying, or by averaging other people's views. Rather, you used your own background knowledge and preferences to put into perspective information you were given about Clinton, and to arrive by a mixture of affective reactions and inferences at your present views. The fact that your views are similar to many other people's may be explained not at all by a copying process, and only partly by an influence process; it may crucially involve the convergence of your affective and cognitive processes with those of many people towards some psychologically attractive type of views in the vast range of possible views on Clinton.

Take languages as a second illustration (see also Boyer 1993: 281). Languages are, at first blush, superb examples of memes: complex skills transmitted from generation to generation and similar enough across individuals to allow communication. However, as Noam Chomsky argued long ago (1972, 1975, 1986), a language such as 'English' is an abstraction to which correspond, in speaker's minds, mental grammars, and, in the environment, linguistic utterances. Individuals never encounter other people's grammar or representations of other people's grammar. Individual learners develop their own grammar on the basis of a large but limited set of linguistic utterances. Different individuals encounter very different sets of utterances. Acquiring a language does not consist in imitating these utterances. In fact, most utterances are never repeated. New utterances are not derived either by averaging or recombining old ones.

Clearly, what happens in language acquisition is that utterances are used as evidence for the construction of a mental grammar. How constraining is this evidence? Chomsky argued – quite convincingly – that the linguistic evidence available to the child vastly underdetermines the grammar. Moreover, many utterances, being grammatically defective, are bad evidence for the grammar to be constructed. Given this underdetermination, and given the differences in the inputs available to different children, the fact that children do each develop a grammar and, moreover, that within the same community these grammars converge, raises a deep problem. Again we owe to Chomsky at least the general form of the solution: there is a domain-specific, genetically specified language acquisition device in every child's mind. In the vast space of possible uses of the stimulation provided by linguistic utterances, children are attracted towards their use as evidence for grammar construction. In the vast space of logically possible grammars, they choose among just a few psychological possibilities, and end up converging on the one grammar psychologically available in the vicinity of the evidence they have been given. Just as it does not matter on which side of the trough you drop the ball, it will roll to the bottom, so it does not matter which French utterances a French child hears, she will construct a French grammar.

As a third example, take 'Little Red Riding Hood', as good an example of a meme as you will ever get. Here, there is no question but that many individuals who hear the tale do aim at retelling it, if not verbatim, at least in a manner faithful to its content. Of course, they don't always succeed, and many of the public versions produced by one teller for the sake of one or a few hearers differ from the several standard versions.

For instance, suppose an incompetent teller has the hunters extract Little Red Riding Hood from the Big Bad Wolf's belly, but forgets the grandmother. Meme theorists might want to argue – and I would agree with them – that such a version is less likely to be replicated than the standard one. The meme theorists' explanation would be that this version is less likely to have descendants. This is indeed plausible. There is another explanation, however, which is also plausible: that hearers whose knowledge of the story derives from this defective version are likely to consciously or unconsciously correct the story when they retell it, and, in their narrative, to bring the grandmother back to life too. In the logical space of possible versions of a tale, some versions have a better form: that is, a form seen as being without either missing or superfluous parts, easier to remember, and more attractive. The factors that make for a good form may be rooted in part in universal human psychology and in part in a local cultural context. In remembering and verbalizing the story, tellers are attracted towards the better forms. Both explanations, in terms of selection and in terms of attraction, may be simultaneously true, then; the reason why defective versions have fewer replicas may be both that they have fewer descendants and that the descendants they have are particularly unlikely to be replicas.

I hope the general idea is now clear: there is much greater slack between descent and similarity in the case of cultural transmission than there is in the biological case. Most cultural descendants are transformations, not replicas. Transformation implies resemblance: the smaller the degree of transformation, the greater the degree of resemblance. But resemblance among cultural items is greater than one would be led to expect by observing actual degrees of transformation in cultural transmission. Resemblance among cultural items is to be explained to some important extent by the fact that transformations tend to be biased in the direction of attractor positions in the space of possibilities.

How, then, should cultural transmission be modelled? Isn't a

Darwinian selection model still the best approximation – to be corrected maybe, but not discarded? To try and answer, let me review the case by means of simple, sketchy formal considerations.

Imagine a population of items that are individually capable of begetting descendants, and that have a limited life-span. Let us imagine that these items come in 100 types, with relationships of similarity among the types such that we may represent the space of possibilities by means of a 10 by 10 matrix (see fig. 1). Imagine some initial stage (which might, e.g., have been experimentally contrived) at which we have a random distribution of, say, 10,000 items among the 100 types. Suppose we examine our population after a number of

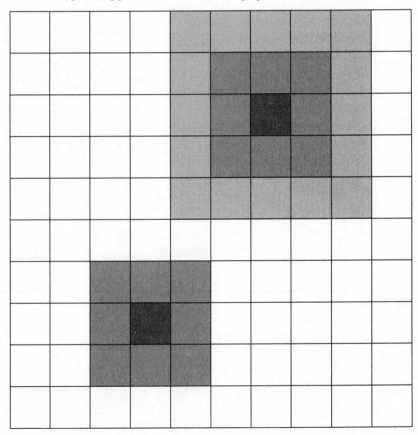

Figure 1 The space of possibilities. After a few generations, density is greater in the two shaded areas.

generations, and observe a different distribution. While the overall size of the population is roughly the same, and there is still a scatter of items across the space of possibilities, some types are now much better represented than others. More specifically, we observe that items tend to be concentrated at and around two types. Imagine that repeated observations show this distribution to be roughly stable.

A well-known kind of explanation of such a state of affairs would be that some of the types were better at replicating initially, and increased in numbers until a kind of ecological equilibrium was reached, at which the more successful types can keep up a higher representation than the others. Suppose, however, that we investigate the manner in which items in this population actually beget descendants, and discover that an offspring is *never* of the same type as its parent! Rather, the offspring is always of one of the eight types adjacent in the matrix to that of its parent (see fig. 2). Ecological equilibrium among differently endowed replicators cannot then be the explanation since this is a transformative, rather than a reproductive, descent system.

An alternative explanation would start from the assumption that the eight possibilities for the offspring of a parent of a given type are not equiprobable. A parent is more likely to beget a transform that differs from it in a given direction. Suppose that the differences in transformation probabilities are such that the matrix has two attractor points.[34] If we trace the descent line of a given item, it will not look like a true random walk in the space of possibilities, but will seem, rather, to be attracted towards one or the other of these attractors (fig. 3) If the departure point of a descent line is far from the attractors, then it is likely that the arrival point will be near one of them. If the departure point is near an attractor, then it is likely that the whole line will stay in its vicinity.

If items begot replicas, then differences in initial reproductive success and ecological equilibrium would explain the observed distribution. Since items beget transforms, however, differences in transformation probabilities provide a better explanation.

Transformation and replication can combine. For instance, all the types might have at all times an equal probability of, say, 1 in 9 of replicating instead of transforming. In this case, although some replication would occur, the difference in distribution between the types

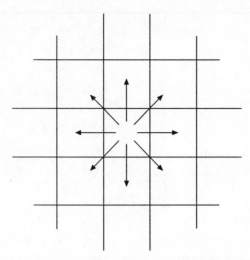

Figure 2 An item begets descendants of neighbouring types.

would be entirely explained by differences in the probabilities of given transformations. Or the probability of an item replicating rather than transforming might differ according to its type. We could, then, in principle, have a dual explanation invoking both reproductive success and attraction. However, in such a case, for the sake of generality and simplicity, reproductive success in a given region is better considered as defining, or contributing to the definition of, that region as an attractor.

The multiplicity and varying number of 'parents', or sources, for the same item, which is, as we noted, a typical aspect of cultural evolution, is also more naturally handled in terms of attraction. The generation of new items in a space of possibilities with attractor regions is to be expected somewhere between existing items and nearby attractors. A simple distance metric is not to be expected, however. The actual mechanisms of generation determine both which items will be transformed and in what manner.

The attraction model easily incorporates the influence model as a special case: the case in which the space of possibility does not include a nearby attractor, and where a simple metric will predict where new items will emerge.

Note that an attractor, as I have characterized it, is an abstract,

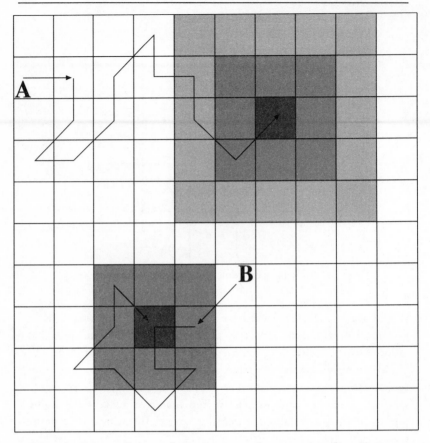

Figure 3   Descent lines tend to move towards an attractor (line A) or to
stay in its vicinity (line B).

statistical construct, like a mutation rate or a transformation probability. To say that there is an attractor is just to say that, in a given space
of possibilities, transformation probabilities form a certain pattern: they
tend to be biased so as to favour transformations in the direction of
some specific point, and therefore cluster at and around that point. An
attractor is not a material thing; it does not physically 'attract' anything. To say that there is an attractor is not to give a causal explanation; it is to put in a certain light what is to be causally explained:
namely, a distribution of items and its evolution, and to suggest the
kind of causal explanation to be sought: namely, the identification of
genuine causal factors that bias micro-transformations.

Ecological and Psychological Factors of Attraction

The existence of attractors is to be explained by two kinds of factors: psychological and ecological. The environment determines the survival and composition of the culture-bearing population; it contains all the inputs to the cognitive systems of the members of the population; it determines when, where and by what medium transmission may occur; it imposes constraints on the formation and stability of different types of public productions. The mental organization of individuals determines which available inputs are processed, how they are processed, and which information guides behaviours that, in turn, modify the environment.

Psychological factors interact with ecological factors at several levels, corresponding to different time-scales: that of biological evolution, that of social and cultural history, that of the cognitive and affective development of individuals, and that of micro-processes of transmission.

It is within the time-scale of biological evolution that a species emerges endowed with mental capacities that make cultural transmission possible. The role of biology is not just to make cultures possible, without any effect on their character and content. The picture of the human mind/brain as a blank slate on which different cultures freely inscribe their own world-view, the picture of world-views as integrated systems wholly determined by socio-cultural history – these pictures, which many social scientists still hold, are incompatible with our current understanding of biology and psychology.

The brain is a complex organ. Its evolution has been determined by the environmental conditions that could enhance or hamper the chances of our ancestors to have offspring throughout phylogeny. There are good reasons to believe that the brain contains many sub-mechanisms, or 'modules', which evolved as adaptations to these environmental opportunities and challenges (Cosmides and Tooby 1987, 1994; Tooby and Cosmides 1989, 1992). Mental modules – that is, adaptations to an ancestral environment – are crucial factors in cultural attraction. They tend to fix a lot of cultural content in and around the cognitive domain the processing of which they specialize in (see chapter 6 below).

Evolutionary pressures are likely to have favoured not only the emergence of specialized mental mechanisms, but also some degree of cognitive efficiency within each of these mechanisms and in their mutual articulation. At any given time, humans perceive more phenomena than they are able to pay attention to, and they have more information stored in memory than they can exploit. Cognitive efficiency involves making the right choices in selecting which available new information to attend to and which available past information to process it with. The right choices in this respect consist in bringing together input and memory information, the joint processing of which will provide as much cognitive effect as possible for as little mental effort as possible.

Deirdre Wilson and I have argued that the effect–effort balance in the processing of any given piece of information determines its degree of relevance (Sperber and Wilson 1986/1995). We claim that human cognitive processes are geared to the maximization of relevance. Most factors of relevance are highly idiosyncratic, and have to do with the individual's unique location in time and space. Some factors of relevance, however, are rooted in genetically determined aspects of human psychology. Thus, the processing of stimuli for which there exists a specialized module requires comparatively less effort and is potentially more relevant. For instance, from birth onwards, humans expect relevance from the sounds of speech (an expectation often disappointed, but hardly ever given up).

It is plausible that individuals should be equipped so as to tend to optimize the effect–effort balance not just on the input side, but also on the output side. Public productions, from bodily movements, to speech, to buildings, even when they are modelled on some previous productions, are likely to move towards forms where the intended effect can be achieved at minimal cost.

Human culture has been around long enough for biological evolution to have been affected by it, in turn. Gene-culture 'coevolution' (Boyd and Richerson 1985; Durham 1991; Lumsden and Wilson 1981) helps explain in particular the existence in humans of abilities that are specifically geared to cultural interaction, such as the language faculty (Pinker and Bloom 1990; Pinker 1994). Gene–culture co-evolution is, however, too slow a process to explain cultural changes in historical time.

Generation after generation, humans are born with essentially the same mental potential. They realize this potential in very diverse ways. This is due to the different environments, and in particular to the different cultural environments, into which they are born. However, from day one, an individual's psychology is enriched and made more specific by cultural inputs. Each individual quickly becomes one of the many loci among which is distributed the pool of cultural representations inhabiting the population. The cultural history of a population is both that of its pool of cultural representations and that of its cultural environment. These macro-ensembles – the pool of representations and the environment – evolve as an effect of micro-processes, where the causes belong to the environment and the effect to the pool, or conversely.

In general, the phrase 'cultural environment' is used very loosely, and refers to a collection of meanings, values, techniques and so forth. So understood, it has little to do with the physical environment. Its ontological status is, at best, very vague; its causal powers are mysterious. By 'cultural environment', I mean an ensemble of material items: all the public productions in the environment that are causes and effects of mental representations. The cultural environment thus understood blends seamlessly with the physical environment of which it is a part. The causal powers it exerts on human minds are unproblematic: public productions affect sense-organs in the usual, material way. They trigger the construction of mental representations the contents of which are partly determined by the properties of the triggering stimuli, and partly by pre-existing mental resources.

Cultural attractors emerge, wane, or move, some rapidly, others slowly, some suddenly, over historical time. Some of these changes have ordinary ecological causes: over-exploited ecological niches lose their economic attraction; rarely walked paths become overgrown; some practices tend to increase, and others to decrease, the size of the populations that might be attracted to them, and so on.

Most historical changes in attractors, however, are to be explained in terms of interactions between ecological and psychological factors of attraction of a kind specific to cultural evolution. The cultural environment causes at every instant the formation of mental representations, some of which themselves cause public productions, and so on. This process modifies the relative density of mental representations, as well as that of public productions, in different areas of the space of possibilities. In particular, density tends to increase in the vicinity of attractors. An increase in the density of public productions in the vicinity of an attractor tends to reinforce the attractor, if only because of the increase in probability that attention will be paid to these more numerous productions. On the other hand, an increase in the density of mental representations in the vicinity of an attractor may weaken the attractor. The repetition of representations having the same tenor may decrease their relevance and bring individuals either to loose interest in them or to reinterpret them differently.

Established practices (in matters of dress, food, etiquette, etc.) act as strong attractors. At the same time, because of their expectability, established practices are often low on relevance, while manifest departures from established practices are often an easy way to attract attention and achieve high relevance. Once public productions massively converge towards some cultural attractor, they may foster the emergence of nearby competing attractors. This is illustrated in a dramatic way by the rapid turnover of fashions, which quickly lose their power because of their very success.

When, on the contrary, one encounters practices that remain stable for generations, one may suppose that they somehow maintain a sufficient level of relevance in spite of repetition, and try to see whether such is indeed the case, and why. A repetitive practice may remain relevant because its effects are. This is the case, for instance, with technological practices, the economic effects of which are important to people's welfare, or even survival. A repetitive practice may remain relevant because it is in competition with other practices, and the choice of one rather than the others by a given individual at a given time may be quite consequential. This is the case with practices used to assert one's belonging to some minority. A repetitive practice may remain relevant because different individuals are in competition for the right to engage in it, and because success in this competition is consequential. This is the case with ritual practices marking promotion to some desired status. A repetitive practice may remain relevant because, without perceptibly modifying its public form, it lends itself to different interpretations according to the agent, the circumstances, and the stage in the life cycle. This possibility of reinterpretation is typical of religious practices (see Sperber 1975b).

On the time-scale of individual life cycles, ecological and psychological factors also interact in a specific manner. At different stages of their psychological development, individuals are attracted in different directions. Initially, the main psychological factors of attraction are genetically determined; but experience – that is, the cognitive effects of past interactions with the environment – becomes an increasingly important factor in attraction.

For much of childhood, information that allows the child to develop competencies for which she has an innate disposition is attended to and used for this purpose. The child becomes a competent speaker, a competent climber, thrower, catcher, eater, drinker, a competent manipulator of objects, a competent recognizer of animals, a competent predictor of other people's behaviour, and so on. In all these domains, new information achieves relevance easily, because it meets the not yet satisfied needs of specialized modules. As basic competencies are acquired, however, attraction shifts to new information relevant in the context of the already acquired basic knowledge. It shifts in particular to cultural information – for instance, religious representations – that seems to challenge basic competencies. It shifts also to information relevant to the various goals that the individual has acquired the ability to conceive and pursue.

The contribution of individuals to cultural transmission varies throughout the lifecycle. Not only do individuals transmit different amounts and different contents, they also transform what they transmit in different directions, and transmit to different audiences according to their stage in life. The amplitude of transformations also varies with the age and social role of the communicator and with those of the audience. In some configurations, a relatively conservative communication appears more relevant. In other configurations, the search for relevance demands innovation. From the point of view of individuals, cultural attractors seem to move along a path that in fact combines historical changes with the individuals' own movement in their life cycle and in their social relationships.

It is the micro-processes of cultural transmission that make possible gene–culture co-evolution, and that bring about the historical evolution of culture and the cultural development of individuals. My argument throughout this book has been that these micro-processes are not, in general, processes of replication. I am not denying that replications occur and play a role in cultural evolution. I am arguing that replications are better seen as limiting cases of transformations. Constructive cognitive processes are involved both in representing cultural inputs and in producing public outputs. All outputs of individual mental processes are influenced by past inputs. Few outputs are mere copies of past inputs. The neo-Darwinian model of culture is based on an idealization, which is good scientific practice. However, this idealization is itself based on a serious distortion of the relevant facts, and this is where the problem lies.

The neo-Darwinian model and the ideas of replication and selection seemed to offer an explanation of the existence and evolution of relatively stable cultural contents. How come, if replication is not the norm, that among all the mental representations and public productions that inhabit a human population and its common environment, it is so easy to discern stable cultural types, such as common views on Bill Clinton, tellings of 'Little Red Riding Hood', English utterances, and also handshakes, funerals and pick-up trucks? For two reasons: first, because, through interpretative mechanisms the mastery of which is part of our social competence, we tend to exaggerate the similarity of cultural tokens and the distinctiveness of types (see ch. 2); and second, because, in forming mental representations and public productions, to some extent all humans, and to a greater extent all members of the same population at any one time, are attracted in the same directions.

Even if it contrasts with the neo-Darwinian models of culture put forward by Dawkins and others, the model of cultural attraction that I have outlined is, quite obviously, of Darwinian inspiration in the way it explains large-scale regularities as the cumulative effect of micro-processes. The culture of a given population is described as a distribution of mental representations and public productions. Cultural evolution is explained as the cumulative effect of differences in frequency between different possible transformations of representations and of productions in the process of transmission. In the study of cultural evolution, borrowing Darwin's selection model is not the only way, and may not be the best way, to take advantage of Darwin's most fundamental insight.[35]

6

Mental Modularity and Cultural Diversity

In *The Modularity of Mind*, published in 1983, Jerry Fodor attacked the then dominant view that there are no important discontinuities between perceptual processes and conceptual processes. Information flows freely, 'up' and 'down', between these two kinds of processes, and beliefs inform perception as much as they are informed by it. Against this view, Fodor argued that perceptual processes (and also linguistic decoding) are carried out by specialized, rather rigid mechanisms. These 'modules' each have their own proprietary data base, and do not draw on information produced by conceptual processes.

Although this was probably not intended and has not been much noticed, *Modularity of Mind* was a paradoxical title, for, according to Fodor, modularity is to be found only at the periphery of the mind, in its input systems.[36] In its centre and bulk, Fodor's mind is decidedly *non*-modular. Conceptual processes – that is, thought proper – are presented as a big holistic lump lacking joints at which to carve. Controversies have focused much more on the thesis that perceptual and linguistic decoding processes are modular, than on the alleged non-modularity of thought.

In this chapter, I have two aims. The first is to defend the view that thought processes might be modular too (what Fodor (1987a:

This chapter is based on my contribution to a conference organized in 1990 in Ann Arbor by Scott Atran, Susan Gelman, Larry Hirschfeld and myself on 'Domain Specificity in Cognition and Culture', My contribution was published under the title 'The Modularity of Thought and the Epidemiology of Representations' in L. B. Hirschfeld and S. A. Gelman (eds), *Mapping the Mind: Domain Specificity in Cognition and Culture* (New York: Cambridge University Press, 1994), 39–67.

27) calls 'modularity theory gone mad' – oh well!). Let me echo
Fodor, however, and say that, 'when I speak of a cognitive system as
modular, I shall ... always mean "to some interesting extent" '
(Fodor 1983: 37). My second aim is to articulate together a modular
view of human thought and the naturalistic view of human culture
that I have been developing under the label 'epidemiology of representations'. These aims are closely related: cultural diversity has
always been taken to show how plastic the human mind is, whereas
the modularity of thought thesis seems to deny that plasticity. I want
to show how, contrary to the received view, organisms endowed
with truly modular minds might engender truly diverse cultures.

Two Common-Sense Arguments Against the
Modularity of Thought

Abstractly and roughly at least, the distinction between perceptual
and conceptual processes is clear. Perceptual processes have, as input,
information provided by sensory receptors and, as output, a conceptual representation categorizing the object perceived. Conceptual
processes have conceptual representations both as input and as output. Thus, seeing a cloud and thinking 'Here is a cloud' is a perceptual process. Inferring from this perception 'It might rain' is a
conceptual process.

The rough idea of modularity is also clear. A cognitive module is
a genetically specified computational device in the mind/brain
(henceforth: the mind) that works pretty much on its own on inputs
pertaining to some specific cognitive domain and provided by other
parts of the nervous systems (e.g. sensory receptors or other modules). Given such notions, the view that perceptual processes might
be modular is indeed quite plausible, as argued by Fodor. On the
other hand, there are two main common-sense arguments (and several more technical ones) that lead one to expect conceptual thought
processes not to be modular.

The first common-sense argument against the modularity of
thought has to do with the integration of information. The conceptual level is the level at which information from different input
modules, each presumably linked to some sensory modality, gets

integrated into a modality-independent medium. A dog can be seen, heard, smelled, touched and talked about: the percepts are different, the concept is the same. As Fodor points out,

> [T]he general form of the argument goes back at least to Aristotle: The representations that input systems deliver have to interface somewhere, and the computational mechanisms that effect the interface must ipso facto have access to information from more than one cognitive domain. (Fodor 1983: 101–2)

The second common-sense argument against the modularity of thought has to do with cultural diversity and novelty. An adult human's conceptual processes range over an indefinite variety of domains, including party politics, baseball history, motorcycle maintenance, Zen Buddhism, French cuisine, Italian opera, chess playing, stamp collecting and Fodor's chosen example, modern science. The appearance of many of these domains in human cognition is very recent, and not relevantly correlated with changes in the human genome. Many of these domains vary dramatically in content from one culture to another, or are not found at all in many cultures. It would be absurd, therefore, to assume that there is an *ad hoc*, genetically specified preparedness for these culturally developed conceptual domains.

These two common-sense arguments are so compelling that Fodor's more technical considerations (having to do with 'isotropy', illusions, rationality, etc.) look like mere nails in the coffin of a dead idea. My goal will be to shake the common-sense picture and to suggest that the challenge of articulating modularity, conceptual integration, and cultural diversity may be met, and will turn out to be a source of psychological and anthropological insights.

Notice, to begin with, that both the informational integration argument and the cultural diversity argument are quite compatible with *partial* modularity at the conceptual level. True, it would be functionally self-defeating to reproduce at the conceptual level the same domain partition found at the perceptual level, and to have a different conceptual module treat separately the output of each perceptual module. No integration whatsoever would take place, and the dog seen and the dog heard could never be one and the same mastiff Goliath. But who says conceptual domains have to match perceptual domains? Why not envisage, at the conceptual level, a wholly different, more or less orthogonal domain partition, with domain-specific conceptual mechanisms each getting their inputs from several input mechanisms? For instance, all the conceptual outputs of perceptual modules that contain the concept MASTIFF (and that are therefore capable of recognizing the presence of a mastiff) might be fed into a specialized module (say a domain-specific inferential device handling living-kind concepts) which takes care, *inter alia* of Goliath *qua* mastiff. Similarly, all the conceptual outputs of input modules which contain the concept THREE might be fed into a specialized module which handles inference about numbers, and so forth. In this way, information from different input devices might get genuinely integrated, though not into a single, but into several conceptual systems.

Of course, if you have, say, a prudential rule that tells you to run away when you encounter more than two bellicose dogs, you would not really be satisfied to be informed by the living-kinds module that the category BELLICOSE DOG is instantiated in your environment, and by the numerical module that there are more than two of something. Some further, at least partial, integration had better take place. It might even be argued – though *that* is by no means obvious – that a plausible model of human cognition should allow for *full* integration of all conceptual information at some level. Either way, partial or full integration might take place further up the line, among the outputs of conceptual, rather than of perceptual, modules. Conceptual integration is not incompatible with at least some conceptual modularity.

Similarly, the conceptual diversity argument implies that some conceptual domains (expertise in postal stamps for instance) could not be modular. It certainly does not imply that none of them could be. Thus, in spite of superficial variations, living-kind classification exhibits strong commonalities across cultures (see Berlin 1978) in a manner that does suggest the presence of a domain-specific cognitive module (see Atran 1987, 1990).

The thesis that some central thought processes might be modular gets support from a wealth of recent work (well illustrated in Hirschfeld and Gelman 1994) tending to show that many basic conceptual thought processes found in every culture and in every fully developed human are governed by domain-specific competences. For instance, it is argued that people's ordinary understanding of the movements of an inert solid object, of the appearance of an organism, or of the actions of a person are based on three distinct mental mechanisms: a naïve physics, a naïve biology and a naïve psychology (see e.g. Atran 1987, 1994; Carey 1985; Keil 1989, 1994; Leslie 1987, 1988, 1994; Spelke 1988). It is argued, moreover, that these mechanisms, at least in rudimentary form, are part of the equipment that makes acquisition of knowledge possible, rather than acquired competences.

Accepting as a possibility some degree of modularity in conceptual systems is innocuous enough. Fodor himself recently considered favourably the view that 'intentional folk psychology is, essentially, an innate, *modularized* database' (1992: 284, my emphasis) without suggesting that he was thereby departing from his former views on modularity. But what about the possibility of *massive* modularity at the conceptual level? Do the two common-sense arguments, integration and diversity, really rule it out?

Modularity and Evolution

If modularity is a genuine natural phenomenon, an aspect of the organization of the brain, then what it consists of is a matter for discovery, not stipulation. Fodor himself discusses a number of characteristic and diagnostic features of modularity. Modules, he argues, are 'domain-specific, innately specified, hardwired, autonomous' (1983: 36). Their operations are mandatory (p. 52) and fast (p. 61). They are 'informationally encapsulated' (p. 64): that is, the only background information available to them is that found in their proprietary data base. They are 'associated with fixed neural architecture' (p. 98). Fodor discusses still other features that are not essential to the present discussion.

There is one feature of modularity that is implied by Fodor's description, although he does not mention or discuss it. If, as Fodor argues, a module is innately specified, hard-wired and autonomous, then it follows that *a cognitive module is an evolved mechanism with a*

distinct phylogenetic history. This is a characteristic, but hardly a diagnostic feature, because we know close to nothing about the actual
evolution of cognitive modules. But I have been convinced by Leda
Cosmides and John Tooby (see Cosmides 1989; Cosmides and
Tooby 1987, 1994; Tooby and Cosmides 1989, 1992)[37] that we
know enough about evolution, on the one hand, and cognition, on
the other, to come up with well-motivated (though, of course, tentative) assumptions as to when to expect modularity, what properties
to expect of modules, and even what modules to expect. This section of the chapter owes much to their ideas.

Fodor himself mentions evolutionary considerations, but only in
passing. He maintains that, phylogenetically, modular input systems
should have preceded non-modular central systems:

> Cognitive evolution would thus have been in the direction of gradu-
> ally freeing certain sorts of problem-solving systems from the con-
> straints under which input analysers labour – hence of producing, as a
> relatively late achievement, the comparatively domain-free inferential
> capacities which apparently mediate the higher flights of cognition.
> (Fodor 1983: 43)

Let us spell out some of the implications of Fodor's evolutionary
suggestion. At an early stage of cognitive evolution we should find
modular sensory input analysers directly connected to modular
motor controllers. There is no level yet where information from
several perceptual processes could be integrated by a conceptual
process. Then there emerges a conceptual device – that is, an inferential device that is not itself directly linked to sensory receptors.
This conceptual device accepts input from two or more perceptual
devices, constructs new representations warranted by these inputs,
and transmits information to motor control mechanisms.

Initially, of course, this conceptual device is just another module.
It is specialized, innately wired, fast, automatic and so forth. Then,
so the story should go, it grows and becomes less specialized; possibly it merges with other similar conceptual devices, to the point
where it is a single big conceptual system, able to process all the outputs of all the perceptual modules, and able to manage all the conceptual information available to the organism. This true central

system cannot, in performing a given cognitive task, activate all the data accessible to it or exploit all of its many procedures. Automaticity and speed are no longer possible. Indeed, if the central system automatically did what it is capable of doing, this would trigger a computational explosion with no end in sight.

An evolutionary account of the emergence of a conceptual module in a mind that has known only perceptual processes is simple enough to imagine. Its demodularization would be much harder to explain.

A toy example might go like this. Organisms of a certain species, call them 'protorgs', are threatened by a danger of a certain kind. This danger (the approach of elephants that might trample on the protorgs, as it might be) is signalled by the co-occurrence of a noise N and soil vibrations V. Protorgs have an acoustic perception module that detects instances of N and a vibration-perception module that detects instances of V. The detection either of N by one perceptual module or of V by the other activates an appropriate flight procedure. Fine, except that when N occurs alone, or when V occurs alone, it so happens that there is no danger. So protorgs end up with a lot of 'false positives', uselessly running away, and thus wasting energy and resources.

Some descendants of the protorgs, call them 'orgs', have evolved another mental device: a conceptual inference mechanism. The perceptual modules no longer directly activate the flight procedure. Rather their relevant outputs – that is, the identification of noise N and that of vibrations V – go to the new device. This conceptual mechanism acts essentially as an AND-gate. When, and only when, both N and V have been perceptually identified, does the conceptual mechanism get into a state that can be said to represent the presence of danger, and it is this state that activates the appropriate flight procedure.

Orgs, so the story goes, competed successfully with protorgs for food resources, and that is why you won't find protorgs around. The orgs' conceptual mechanism, though not an *input* module, is nevertheless a clear case of a module: it is a domain-specific problem-solver; it is fast, informationally encapsulated, associated with fixed neural architecture, and so forth. Of course, it is a tiny module, but nothing stops us from imagining it becoming larger. Instead of accepting just two bits of information from two simple perceptual modules, the conceptual module could come to handle more information from more sources, and to control more than a single motor procedure, but still be domain-specific, automatic, fast and so on.

At this juncture, we have two diverging evolutionary scenarios on offer. According to the scenario suggested by Fodor, the conceptual module should evolve towards reduced domain specificity, less informational encapsulation, less speed and so on. In other words, it should become less and less modular, possibly merge with other demodularized devices, and end up like the kind of central system with which Fodor believes we are endowed ('Quineian', 'isotropic', etc.). There are two gaps in this scenario. The first has to do with mental mechanisms, and is highlighted by Fodor himself in his 'First Law of the Nonexistence of Cognitive Science'. This law says in substance that the mechanisms of non-modular thought processes are too complex to be understood. So, just accept that there are such mechanisms, and don't ask how they work!

The second gap in Fodor's scenario has to do with the evolutionary process that is supposed to bring about the development of such a mysterious mechanism. No doubt, it might be advantageous to trade a few domain-specific inferential micro-modules for an advanced all-purpose macro-intelligence, if there is such a thing. For instance, super-orgs endowed with general intelligence might develop technologies to eradicate the danger once and for all, instead of having to flee again and again. But evolution does not offer such starkly contrasting choices. The available alternatives at any one time are all small departures from the existing state. Selection, the main force driving evolution, is near-sighted (whereas the other forces – genetic drift, etc. – are blind). An alternative that is immediately advantageous is likely to be selected from the narrow available range, and this may bar the path to highly advantageous long-term outcomes. A demodularization scenario is implausible for this very reason.

Suppose, indeed, that the conceptual danger analyser is modified in some mutant orgs, not in the direction of performing better at its special task, but in that of less domain specificity. The modified conceptual device processes not just information relevant to the orgs' immediate chances of escape, but also information about innocuous features of the dangerous situation, and about a variety of innocuous situations exhibiting these further features; the device draws inferences not just of an urgent practical kind, but also of a more theoretical character. When danger is detected, the new, less modular system does not automatically trigger flight behaviour, and when it does, it does so more slowly – automaticity and speed go with modularity – but it has interesting thoughts that are filed in memory for the future ... if there *is* any future for mutant orgs endowed with this partly demodularized device.

Of course, speed and automaticity are particularly important for danger analysers, but less so for other plausible modules – for instance, modules governing the choice of sexual partners. However, the general point remains: evolved cognitive modules are likely to be answers to specific, usually environmental, problems. Loosening the domain of a module will bring about, not greater flexibility, but greater slack in the organism's response to the problem. To the limited extent that evolution tends toward improving a species' biological endowments, then we should generally expect improvements in the manner in which existing modules perform their task, the emergence of new modules to handle other problems, but not demodularization.

True, it is possible to conceive of situations in which the marginal demodularization of a conceptual device might be advantageous, or at least not detrimental, in spite of the loss of speed and reliability involved. Imagine, for instance, that the danger the conceptual module was initially selected to analyse has vanished from the environment; then the module is not adapted any more, and a despecialization would do no harm. On the other hand, why should it do any good? Such odd possibilities fall well short of suggesting a positive account of the manner in which, to repeat Fodor's words, 'cognitive evolution would ... have been in the direction of gradually freeing certain sorts of problem-solving systems from the constraints under which input analysers labour'. It is not that this claim could not be right, but it is poorly supported. In fact the only motivation for it seems to be the wish to integrate the belief that human thought processes are non-modular into some evolutionary perspective, however vague. Better to render official the explanatory gap with a 'Second Law of the Nonexistence of Cognitive Science', according to which the forces that have driven cognitive evolution can never be identified.[38] Just accept that cognitive evolution occurred (and resulted in the demodularization of thought), and don't ask how.

Instead of starting from an avowedly enigmatic view of *Homo sapiens*'s thought processes and concluding that their past evolution is an unfathomable mystery, one might start from evolutionary considerations plausible in their own right and wonder what kind of cognitive organization these might lead one to expect in a species which we know to rely heavily on its cognitive abilities for its survival. This yields our second scenario.

As already suggested, it is reasonable to expect conceptual modules to gain in complexity, fine-grainedness and inferential sophistication *in the performance of their function*. As with any biological device, the function of a module may vary over time; but there is no reason to expect new functions to be systematically more general than old ones. It is reasonable, on the other hand, to expect new conceptual modules to appear in response to different kinds of problems or opportunities. Thus more and more modules might accumulate.

Because cognitive modules are each the result of a different phylogenetic history, there is no reason to expect them all to be built on the same general pattern and elegantly interconnected. Though most, if not all, conceptual modules are inferential devices, the inferential procedures that they use may be quite diverse. Therefore, from a modular point of view, it is unreasonable to ask about the general form of human inference (logical rules, pragmatic schemas, mental models, etc.) as is generally done in the literature on human reasoning (see Manktelow and Over 1990 for a recent review).

The 'domains' of modules may vary in character and in size: there is no reason to expect domain-specific modules to each handle a domain of comparable size. In particular, there is no reason to exclude micro-modules the domain of which is the size of a concept, rather than that of a semantic field. In fact, I will argue that many human concepts are individually modular. Because conceptual modules are likely to be many, their interconnections and their connections with perceptual and motor control modules may be quite diverse too. As argued by Andy Clark (1987, 1990), we had better think of the mind as kludge, with sundry bits and components added at different times, and interconnected in ways that would make an engineer cringe.

Modularity and Conceptual Integration

The input to the first conceptual modules to have appeared in cognitive evolution must have come from the perceptual modules. However, once some conceptual modules were in place, their output could serve as input to other conceptual modules.

Suppose the orgs can communicate among themselves by means of a small repertoire of vocal signals. Suppose, further, that the optimal interpretation of some of these signals is sensitive to contextual factors. For instance, an ambiguous danger signal indicates the presence of a snake when emitted by an org on a tree, and approaching elephants when emitted by an org on the ground. Identifying the signals and the relevant contextual information is done by perceptual modules. The relevant output of these perceptual modules is processed by an *ad hoc* conceptual module that interprets the ambiguous signals. Now, it would be a significant improvement if the conceptual module specialized in inferring the approach of elephants would accept as input not only perceptual information on specific noises and soil vibrations, but also interpretations of the relevant signals emitted by other orgs. Then, this danger–inferring conceptual module would receive input not just from perceptual modules but also from another conceptual module, the context-sensitive signal interpreter.

In the human case, it is generally taken for granted that domain-specific abilities can process not just primary information belonging to their domain and provided by perception, but also verbally or pictorially communicated information. Thus experiments on the development of zoological knowledge use as material, not actual animals, but pictures or verbal descriptions. This methodology deserves discussion, but it does not seem to raise serious problems. This in itself is quite remarkable.

Then, too, some conceptual modules might get *all* their input from other conceptual modules. Imagine, for instance, that an org emits a danger signal only when two conditions are fulfilled: it has inferred the presence of a danger, on the one hand, and that of friendly orgs at risk, on the other hand. Both inferences are drawn by conceptual modules. If so, then the conceptual module that decides whether or not to emit the danger signal gets all its input from other conceptual modules, and none from perceptual ones.

We are now envisaging a complex network of conceptual modules. Some conceptual modules get all their input from perceptual modules, others get at least some of their input from conceptual modules, and so forth. Every piece of information may get combined with many others across or within levels and in various ways (though overall conceptual integration seems excluded). What would be the behaviour of an organism endowed with such complex modular thought processes? Surely, we don't know. Would it behave in a flexible manner like humans do? Its responses could at least be extremely fine-grained. Is there more to flexibility than this fine-grainedness? 'Flexibility' is a metaphor without a clear literal interpretation, and therefore it is hard to tell. Still, when we think of flexibility in the human case, we particularly have in mind the ability to learn from experience. Can a fully modular system learn?

Imprinting is a very simple form of modular learning. What, for instance, do orgs know about one another? If orgs are non-learning animals, they might merely be endowed with a conspecific detector and detectors for some properties of other orgs such as sex or age, but might otherwise be unable to detect any single individual as such, not even, say, their own mothers. Or, if they are very primitive learners, they might have a mother detector module the workings of which will be fixed once and for all by the new-born org's first perception of a large moving creature in its immediate vicinity (hopefully its real mother), and of the resulting imprinting of the relevant information. The module then becomes a detector for the particular individual who caused the imprinting.

More generally, I wish to introduce here a technical notion, that of 'initialization', borrowed from computer vocabulary. A cognitive module may, just like a computer program, be incomplete in the sense that specific pieces of information must be fixed before it can function normally. An e-mail program, for instance, may ask you to fix a few parameters (e.g. baud rate or parity) and to fill empty slots (e.g. phone numbers to call or a password). It is only after having been so initialized that your program can work. Similarly, first language learning according to Noam Chomsky (1986) (whose work was seminal in the development of a modularist approach to the human mind – see Hirschfeld and Gelman 1984: introduction) involves, in particular, fixing, for several grammatical parameters common to all languages, the values these parameters take in the language to be learnt, and filling in a lexicon. To initialize a cognitive module is thus a matter of fixing the values of parameters and of filling empty slots. The initialization of the mother detector described in the preceding paragraph involves just filling its single empty slot with the perceptual representation of a single individual.

If they are slightly more sophisticated learners, orgs may have the capacity to construct several detectors for different individual conspecifics. They might have a template module quite similar to a mother detector, except that it can be 'initialized' several times, each time projecting a differently initialized copy of itself that is specialized for the identification of a different individual. Would the initialized copies of the template module be modules too? I don't see why not. The only major difference is that these numerous projected modules seem less likely to be hard-wired than a single mother detector module. Otherwise, both kinds of modules get initialized and operate in exactly the same manner. Of our more sophisticated orgs, we would want to say, then, that they had a modular, domain-specific ability to mentally represent conspecific individuals, an ability resulting in the generation of micro-modules for each individual represented.

Consider in this light the human domain-specific ability to categorize living kinds. One possibility is that there is an initial template module for living-kind concepts that gets initialized many times, producing each time a new micro-module corresponding to one living-kind concept (the DOG module, the CAT module, the GOLDFISH module, etc.) Thinking of such concepts as modules may take some getting used to, I admit. Let me help: concepts are domain-specific (obviously), they have a proprietary data base (the encyclopaedic information filed under the concept), and they are autonomous computational devices (they work, I will argue, on representations in which the right concept occurs, just as digestive enzymes work on food in which the right molecule occurs). When, on top of all that, concepts are in part genetically specified (via some domain-specific conceptual template), they are modular at least to some interesting extent, no?

The template–copy relationship might sometimes involve more levels. A general living-kinds-categorization meta-template could project, not directly concepts, but other, more specific templates for different domains of living kinds. For instance, a fundamental parameter to be fixed might concern the contrast between self-propelled and non-self-propelled objects (Premack 1990), yielding two templates, one for zoological concepts, another for botanical concepts.

Another possibility is that the initial meta-template has three types of features: (1) fixed features that characterize living kinds in general – for instance, it might be an unalterable part of any living-kind concept that the kind is taken to have an underlying essence (Atran 1987; Gelman and Coley 1991; Gelman and Markman 1986, 1987; Keil 1989; Medin and Ortony 1989); (2) parameters with default values that can be altered in copies of the template – for instance, 'self-propelled' and 'non-human' might be revisable features of the initial template; (3) empty slots for information about individual kinds. If so, the default-value template could serve as such for non-human animal concepts. To use the template for plant concepts, or to include humans in a taxonomy of animals, would involve changing a default value of the initial template.

How is the flow of information among modules actually governed? Is there a regulating device? Is it a pandemonium? A market economy? Many types of models can be entertained. Here is a simple possibility.

The output of perceptual and conceptual modules is in the form of conceptual representations. Perceptual modules categorize distal stimuli (things seen, e.g.), and must each have, therefore, the conceptual repertoire needed for the output categorizations of which they are capable. Conceptual modules may infer new output categorizations from the input conceptual representations they process; they must have an input and an output conceptual repertoire to do so. Let us assume that conceptual modules accept as input any conceptual representation in which a concept belonging to their input repertoire occurs. In particular, single-concept micro-modules process all and only representations in which their very own concept occurs. These micro-modules generate transformations of the input representation by replacing the concept with some inferentially warranted expansion of it. They are otherwise blind to the other conceptual properties of the representations they process (in the manner of the 'calculate' procedure in some word processor, which scans the text but 'sees' only numbers and mathematical signs). Generally, the presence of specific concepts in a representation determines what modules will be activated and what inferential processes will take place (see Sperber and Wilson 1986: ch. 2).

A key feature of modularity in Fodor's description is informational encapsulation. A full-fledged module uses a limited data base, and is not able to take advantage of information relevant to its task if that information is in some other data base. Central processes, on the other hand, are not so constrained. On the contrary, they are characterized, by free flow of information. Thus beliefs about Camembert cheese might play a role in forming conclusions about quarks, even though they hardly belong to the same conceptual domain. This last is a fact, and I wouldn't dream of denying it. But what does it imply regarding the modularity of conceptual processes? It implies that one particular modular picture cannot be right. Imagine a single layer of a few large, mutually unconnected modules; then information treated by one module won't find its way to another. If, on the other hand, the output of one conceptual module can serve as input to another one, modules can each be informationally encapsulated, while chains of inference can take a conceptual premiss from one module to the next, and therefore integrate the contribution of each in some final conclusion. A holistic effect need not be the outcome of a holistic procedure.

Once a certain level of complexity in modular thought is reached, modules can emerge whose function is to handle problems raised, not externally by the environment, but internally by the workings of the mind itself. One problem that a rich modular system of the kind we are envisaging would encounter as surely as Fodor's non-modular central processes is the risk of computational explosion.

Assume that a device has emerged, the function of which is to put up on the board, so to speak, some limited information for actual processing. Call this device 'attention'. Think of it as a temporary buffer. Only representations stored in that buffer are processed (by the modules whose input conditions they satisfy), and they are processed only as long as they stay in the buffer. There is, so to speak, competition among representations for attention. The competition tends to work so as to maximize cognitive efficiency; that is, it tends to select for a place in the buffer, and thus for inferential processing, the most relevant information available at the time. There is a much longer story to be told: read *Relevance* (Sperber and Wilson 1986).

Attention is, of course, not domain-specific. On the other hand, it is a clear adaptation to an internal processing problem: the problem encountered by any cognitive system able to identify perceptually and hold in memory much more information than it can simultaneously process conceptually. Such a system must be endowed with a means of selecting the information to be conceptually processed. Relevance-guided attention is such a means. Whether or not it should be called a module does not really matter: attention so conceived fits snugly into a modular picture of thought.

I don't expect these speculations to be convincing – I am only half convinced myself, though I will be a bit more so by the end of this chapter – but I hope they are intelligible. If so, this means that one can imagine a richly modular conceptual system that integrates information in so many partial ways that it is not obvious any more that we, human beings, genuinely integrate it in any fuller way. The argument against the modularity of thought based on the alleged impossibility of modular integration should lose at least its immediate common-sense appeal.

Actual and Proper Domains of Modules

Modules are domain-specific, and many – possibly most – domains of modern human thought are too novel and too variable to be the specific domain of a genetically specified module. This second common-sense argument against the modularity of thought is reinforced by adaptationist considerations: in many domains, cultural expertise is hard to see as a biological adaptation. This is true not just of new domains such as chess, but also of old domains such as music. Expertise in these domains is unlikely, therefore, to be based on an *ad hoc* evolved mechanism. Of course, one can always try to concoct some story showing that, say, musical competence is a biological adaptation. However, merely assuming the adaptive character of a trait without a plausible demonstration is an all too typical misuse of the evolutionary approach.

Let me try an altogether different line. An adaptation is, generally, an adaptation to given environmental conditions. If you look at an adaptive feature just by itself, inside the organism, and forget altogether what you know about the environment and its history, you cannot tell what its function is, what it is an adaptation to. The function of a giraffe's long neck is to help it eat from trees, but in another environment — make it on another planet, to free your imagination — the function of an identical body part on an identical organism could be to allow the animal to see further, or to avoid breathing foul air near the ground, or to fool giant predators into believing that its flesh is poisonous.

A very similar point — or, arguably, a special application of the same point — has been at the centre of major recent debates in the philosophy of language and mind between 'individualists' and 'externalists'. Individualists hold that the content of a concept is in the head of the thinker, or, in other words, that a conceptual content is an intrinsic property of the thinker's brain state. Externalists maintain — rightly, I believe — that the same brain state that realizes a given concept might realize a different concept in another environment, just as internally identical biological features might have different functions in different environments.[39]

The content of a concept is not an intrinsic, but a relational, property[40] of the neural realizer of that concept, and is contingent upon the environment and the history (including the phylogenetic prehistory) of that neural object. This extends straightforwardly to the case of domain-specific modules. A domain is semantically defined, that is, by a concept under which objects in the domain are supposed to fall. The domain of a module is therefore not a property of its internal structure (whether described in neurological or in computational terms).

There is no way in which a specialized cognitive module might pick its domain just by virtue of its internal structure, or even by virtue of its connections to other cognitive modules. All that the internal structure provides is, to borrow an apt phrase from Frank Keil (1994), a *mode of construal*, a disposition to organize information in a certain manner and to perform computations of a certain form. A cognitive module also has structural relations to other mental devices with which it interacts. This determines, in particular, its *input conditions*: through which other devices the information must come, and how it must be categorized by these other devices. But, as long as one remains within the mind and ignores the connections of perceptual modules with the environment, knowledge of the brain-internal connections of a specialized cognitive module does not determine its domain.

The fact that the mode of construal afforded by a mental module might fit many domains does *not* make the module any less domain-specific, just as the fact that my key might fit many locks does not make it any less the key to my door. The mode of construal and the domain, just as my key and my lock, have a long common history. How, then, do interactions with the environment over time determine the domain of a cognitive module? To answer this question, we need to distinguish between the *actual* and the *proper* domain of a module.

The *actual domain* of a conceptual module is all the information in the organism's environment that may (once processed by perceptual modules, and possibly by other conceptual modules) satisfy the module's input conditions. Its *proper domain* is all the information that it is the module's biological function to process. Very roughly, the function of a biological device is a class of effects of that device that contributes to making the device a permanent feature of a viable species. The function of a module is to process a specific range of information in a specific manner. That processing contributes to the reproductive success of the organism. The range of information that it is the function of a module to process constitutes its proper domain. What a module actually processes is information found in its actual domain, whether it also belongs to its proper domain or not.

Back to the orgs. The characteristic danger that threatened them initially was being trampled by elephants. Thanks to a module, the orgs reacted selectively to various signs normally produced, in their environment, by approaching elephants. Of course, approaching elephants were sometimes missed, and other, unrelated and innocuous events sometimes activated the module. But even though the module failed to pick out all and only approaching elephants, we describe its function as having been to do just that (rather than doing what it actually did). Why? Because it is its relative success at that task that explains its having been a permanent feature of a viable species. Even though they were not exactly coextensive, the actual domain of the module overlapped well enough with the approaching elephants domain. Only the latter, however, was the proper domain of the module.

Many generations later, elephants had vanished from the orgs' habitat, while hippopotamuses had multiplied, and now *they* trampled absent-minded orgs. The same module that had reacted to most approaching elephants and a few other, sundry events now reacted to most approaching hippos and a few other, sundry events. Had the module's proper domain become that of approaching hippos? Yes, and for the same reasons as before: its relative success at reacting to approaching hippos explains why this module remained a permanent feature of a viable species.[41]

Today, however, hippopotamuses too have vanished, and there is a railway passing through the orgs' territory. Because orgs don't go near the rails, trains are no danger. However, the same module that had reacted selectively to approaching elephants and then to approaching hippos now reacts to approaching trains (and produces a useless panic in the orgs). The *actual* domain of the module includes mostly approaching trains. Has its *proper* domain therefore become that of approaching trains? The answer should be 'no' this time: reacting to trains is what it does, not its function. The module's reacting to trains does not explain its remaining a permanent feature of the species. In fact, if the module and the species survive, it is in spite of this marginally harmful effect.[42]

Still, an animal psychologist studying orgs today might well come to the conclusion that they have a domain-specific ability to react to trains. She might wonder how they have developed such an ability, given that trains have been introduced into the area too recently to allow the emergence of a specific biological adaptation (the adaptive value of which would be mysterious anyhow). The truth, of course, is that the earlier proper domains of the module, approaching elephants and then hippos, are now empty; that its actual domain is, by accident, roughly coextensive with the set of approaching trains; and that the explanation of this accident is the fact that the input conditions of the module, which had been positively selected in a different environment, happen to be satisfied by trains and hardly anything else in the orgs' present environment.

Enough of toy examples. In the real world, you are unlikely to get elephants neatly replaced by hippos and hippos by trains, and to have each kind in turn satisfying the input conditions of some specialized module. Natural environments, and therefore cognitive functions, are relatively stable. Small shifts of cognitive function are more likely to occur than radical changes. When major changes occur in the environment – for instance, as the result of a natural cataclysm – some cognitive functions are likely just to be lost. If elephants go, so does the function of your erstwhile elephant detector. If a module loses its function, or, equivalently, if its proper domain becomes empty, then it is unlikely that its actual domain will be neatly filled by objects all falling under a single category, such as passing trains. More probably, the range of stimuli causing the module to react will end up being such a medley as to discourage any temptation to describe the actual domain of the module in terms of a specific category. Actual domains are usually not conceptual domains.

Cultural Domains and the Epidemiology of Representations

Most animals get only highly predictable kinds of information from their conspecifics, and not much of it at that. They depend, therefore, on the rest of the environment for their scant intellectual kicks. Humans are special. They are naturally massive producers, transmitters and consumers of information. They get a considerable amount and variety of information from fellow humans, and they even produce and store some for their own private consumption. As a result,

I will argue, the actual domain of human cognitive modules is likely to have become much larger than their proper domain. Moreover, these actual domains, far from being uncategorizable chaos, are likely to be partly organized and categorized by humans themselves. So much so, I will argue, that we should distinguish the *cultural domains* of modules from both their proper and their actual domains.

Just a quick illustration before I give a more systematic sketch and a couple of more serious examples: here is an infant in her cradle, endowed with a domain-specific, modular, naïve physics. The proper domain of that module is a range of physical events that typically occur in nature, the understanding of which will be crucial to the organism's survival. Presumably, other primates are endowed with a similar module. The naïve physics module of the infant chimp (and of the infant Pleistocene *Homo* not yet *sapiens*) reacts to the odd fruit or twig falling, to the banana peel being thrown away, to occasional effects of its own movement, and it may be challenged by the irregular fall of a leaf. Our human infant's module, on the other hand, is stimulated not just by physical events happening incidentally, but also by an 'activity centre' fixed to the side of her cradle, a musical merry-go-round just above her head, balls bounced by older siblings, moving pictures on a television screen, and a variety of educational toys devised to stimulate her native interest in physical processes.

What makes the human case special? Humans change their own environment at a rate that natural selection cannot follow; so, many genetically specified traits of the human organism are likely to be adaptations to features of the environment that have ceased to exist or have greatly changed. This may be true not just of adaptations to the non-human environment, but also of adaptations to earlier stages of the hominid social environment.

In particular, the actual domain of *any* human cognitive module is unlikely to be even approximately coextensive with its proper domain. On the contrary, the actual domain of any human cognitive module is sure, on the contrary, to include a large amount of cultural information that meets its input conditions. This results neither from accident nor from design. It results from a process of social distribution of information.

In the epidemiological perspective advocated in this book, all the information that humans introduce into their common environment can be seen as competing[43] for private and public space and time – that is, for attention, internal memory, transmission and external storage. Many factors affect the chances of some information being successful and reaching a wide and lasting level of distribution, of being stabilized in a culture. Some of these factors are psychological, others are ecological. Most of them are relatively local, others are quite general. The most general psychological factor affecting the distribution of information is its compatibility and fit with human cognitive organization.

In particular, relevant information the relevance of which is relatively independent of the immediate context is, *ceteris paribus*, more likely to reach a cultural level of distribution. Relevance provides the motivation both for storing and for transmitting the information, and independence of an immediate context means that relevance will be maintained in spite of changes of local circumstances – that is, it will be maintained on a social scale. Relevance is always relative to a context however; independence of the immediate context means relevance in a wider context of stable beliefs and expectations. On a modular view of conceptual processes, these beliefs, which are stable across a population, are those which play a central role in the modular organization and processing of knowledge. Thus information that either enriches or contradicts these basic modular beliefs stands a greater chance of cultural success.

I have argued elsewhere (Sperber 1975b, 1980) and in chapters 3 and 4 that beliefs which violate head-on module-based expectations (e.g. beliefs in supernatural beings capable of action at a distance, ubiquity, metamorphosis, etc.) thereby gain a salience and relevance that contribute to their cultural robustness. Pascal Boyer (1990) has rightly stressed that these violations of intuitive expectations in the description of supernatural beings are in fact few, and take place against a background of satisfied modular expectations. Kelly and Keil (1985) have shown that cultural exploitation of representations of metamorphoses are closely constrained by domain-based conceptual structure. Generally speaking, we should expect many culturally successful representations to be squarely grounded in a conceptual module, and at the same time to differ enough from the information found in the module's proper domain to command attention.

A cognitive module stimulates in every culture the production and distribution of a wide array of information that meets its input conditions. This information, being artefactually produced or organized by the people themselves, is from the start conceptualized, and therefore belongs to conceptual domains that I propose to call the module's *cultural domain(s)*. In other words, cultural transmission causes, in the actual domain of any cognitive module, a proliferation of parasitic information that mimics the module's proper domain.

Let me first illustrate this epidemiological approach with speculations on a non-conceptual case, that of music. This is intended to be an example of a way of thinking suggested by the epidemiological approach, rather than a serious scientific hypothesis, which I would not have the competence to develop.

Imagine that the ability and propensity to pay attention to, and analyse, certain complex sound patterns became a factor of reproductive success for a long enough period in human prehistory. The sound patterns would have been discriminable by pitch variation and rhythm. What sounds would have exhibited such patterns? The possibility that springs to mind is human vocal communicative sounds. It need not be the sounds of *Homo sapiens* speech, though. One may imagine a human ancestor with much poorer articulatory abilities, relying more than modern humans do on rhythm and pitch for the production of vocal signals. In such conditions, a specialized cognitive module might well have evolved.

This module would have had to combine the necessary discriminative ability with a motivational force to cause individuals to attend to the relevant sound patterns. The motivation would have to be on the hedonistic side: pleasure and hopeful expectation, rather than pain and fear. Suppose that the relevant sound pattern co-occurred with noise from which it was hard to discriminate it. The human ancestor's vocal abilities may have been quite poor, and the intended sound pattern may have been embedded in a stream of parasitic sounds (a bit like when you speak with a sore throat, a cold, and food in your mouth). Then the motivational component of the module should have been tuned so that detecting a low level of the property suffices to procure a significant reward.

The proper domain of the module we are imagining is the acoustic properties of early human vocal communications. It could be that this proper domain is now empty: another adaptation, the improved modern human vocal tract, may have rendered it obsolete. Or it may be that the relevant acoustic properties still play a role in modern human speech (in tonal languages in particular), so the module is still functional. The sounds that the module analyses, thereby causing pleasure to the organism of which it is a part – that is, the sounds meeting the module's input conditions – are not often found in nature (with the obvious exception of bird-songs). However, such sounds can be produced artificially. And they have been, providing this module with a particularly rich cultural domain: music. The relevant acoustic pattern of music is much more detectable and delectable than that of any sound in the module's proper domain. The reward mechanism, which was naturally tuned for a hard-to-discriminate input, is now being stimulated to a degree that makes the whole experience utterly addictive.

The idea, then, is that humans have created a cultural domain, music, which is parasitic on a cognitive module the proper domain of which pre-existed music and had nothing to do with it. The existence of this cognitive module has favoured the spreading, stabilization and progressive diversification and growth of a repertoire meeting its input conditions. First, pleasing sounds were serendipitously discovered, then sound patterns were deliberately produced and became music proper. These bits of culture compete for mental and public space and time, and ultimately for the chance to stimulate the module in question in as many individuals as possible for as long as possible. In this competition, some pieces of music do well, at least for a time, whereas others are eliminated, and thus music, and musical competence, evolve.

In the case of music, the cultural domain of the module is much more developed and salient than its proper domain, assuming that it still has a proper domain. So much so that it is the existence of the cultural domain and the domain specificity of the competences it manifestly evokes that justifies looking, in the present or in the past, for a proper domain that is not immediately manifest.

In other cases, the existence of a proper domain is at least as immediately manifest as that of a cultural one. Consider zoological knowledge. The existence of a domain-specific competence in the matter is not hard to admit, if the general idea of domain specificity is accepted at all. One way to think of it, as I have suggested, is to suppose that humans have a modular template for constructing concepts of animals. The biological function of this module is to provide humans with ways of categorizing animals they may encounter in their environment and of organizing the information they may gather about them. The proper domain of this modular ability is the living local fauna. What happens, however, is that you end up, thanks to cultural input, constructing concepts for animal species with which you will never interact. If you are a twentieth-century Westerner, you may, for instance, have a cultural sub-domain of dinosaurs. You may even be a dinosaur expert. In another culture, you might have been a dragon expert.

This invasion of the actual domain of a conceptual module by cultural information occurs irrespective of the size of the module. Consider a micro-module such as the concept of a particular animal, say the rat. Again, you are likely to have fixed, in the data base of that module, culturally transmitted information about rats, whether of a folkloristic or of a scientific character, that goes well beyond the proper domain of that micro-module – that is, well beyond information derivable from, and relevant to, interactions with rats. Of course, this cultural information about rats may be of use for your interactions with other human beings, by providing, for instance, a data base exploitable in metaphorical communication.

On the macro–modular side of things, accept for the sake of this discussion that the modular template on which zoological concepts are constructed is itself an initialized version (maybe the default version) of a more abstract living-kinds meta-template. That meta-template is initialized in other ways for other domains (e.g. botany), projecting several domain-specific templates, as I have suggested above. What determines a new initialization is the presence of information that (1) meets the general input conditions specified in the meta-template, but (2) does not meet the more specific conditions found in the already initialized templates. That information need not be in the proper domain of the meta-template module. In other words, the meta-template might get initialized in a manner that fits no proper domain at all, but only a cultural domain. A cultural domain that springs to mind in this context is that of representations of supernatural beings (see Boyer, 1990, 1993, 1994). But there may also be less apparent cases.

Consider in this light the problem raised by Hirschfeld (1988, 1993, 1994). Children, he shows, are disposed to categorize humans into 'racial' groups. Moreover, they draw inferences from these categorizations, as if different racial groups had different 'essences', or 'natures', comparable to the different natures attributed to different animal species. Do children possess a competence the function of which is to develop such categorizations? In other terms, are humans naturally disposed to racism? Avoiding such an unappealing conclusion, it has been suggested (Atran 1990; Boyer 1990) that children transfer to the social sphere a competence that they first developed for natural living kinds, and that they do so in order to make sense of systematic differences in human appearance (e.g. skin colour) that they may have observed. However, Hirschfeld's experimental evidence shows that racial categorization develops without initially drawing on perceptually relevant input. This seems to suggest that there is, after all, a domain-specific competence for racial classification.

What the epidemiological approach suggests is that racial classification might result from a domain-specific, but not innate, template derived from the living-kinds meta-template, through an initialization process triggered by a cultural input. Indeed, recent experiments suggest that, in certain conditions, the mere encounter with a nominal label used to designate a living thing is enough to tilt the child's categorization of that thing toward an 'essentialist' construal, according to which the perceptible features of the species are manifestations of an underlying essence (Markman and Hutchinson 1984; Markman 1990; Davidson and Gelman 1990; Gelman and Coley 1991). It is quite possible, then, that being presented with nominal labels for otherwise undefined and undescribed humans is enough (given an appropriate context) to activate the initialization of the *ad hoc* template. If so, then perception of physical differences among humans is indeed not the triggering factor in racial classification.

There is, as Hirschfeld suggested, a genetically specified competence that determines racial classification without importing its models from another concrete domain. However, the underlying competence need not have racial classification as its proper domain. Racial classification may be a mere cultural domain, based on an underlying competence that does not have any proper domain. The initialization of an *ad hoc* template for racial classification could well be the effect of parasitic, cultural input information on the higher-level learning module, the function of which is to generate *ad hoc* templates for genuine living-kind domains such as zoology and botany. If this hypothesis is correct – mind you, I am not claiming that it is, merely that it may be – then no racist disposition has been selected *for* (Sober 1984) in humans. However, the dispositions that *have* been selected for make humans all too easily susceptible to racism, given minimal, innocuous-looking cultural input.

The relationship between the proper and the cultural domains of the same module is not one of transfer. The module itself does not have a preference between the two kinds of domains, and indeed is blind to a distinction that is grounded in ecology and history. Even when an evolutionary and epidemiological perspective is adopted, the distinction between the proper and the cultural domain of a module is not always easy to draw. Proper and cultural domains may overlap. Moreover, because cultural domains are things of this world, it can be a function of a module to handle a cultural domain, which *ipso facto* becomes a proper domain.

Note that the very existence of a cultural domain is an effect of the existence of a module. Therefore, initially at least, a module cannot be an adaptation to its own cultural domain. It must have been selected because of a pre-existing proper domain. In principle, it might *become* a function of the module to handle its own cultural domain. This would be the case when the ability of the module to handle its cultural domain contributed to its remaining a permanent feature of a viable species. The only clear case of an adaptation of a module to its own effects is that of the linguistic faculty. The linguistic faculty in its initial form cannot have been an adaptation to a public language that could not exist without it. On the other hand, it seems hard to doubt that language has become the proper domain of the language faculty.[44]

If there are modular abilities to engage in specific forms of social interaction (as claimed by Cosmides (1989)), then, as in the case of the language faculty, the cultural domains of these abilities should at least overlap with their proper domains. Another interesting issue in this context is the relationship between numerosity – the proper domain of a cognitive module – and numeracy, an obviously cultural domain dependent on language (see Gelman and Gallistel

1978; Gallistel and Gelman 1992; Dehaene 1992). In general, however, there is no reason to expect the production and maintenance of cultural domains to be a biological function of all, or even most, human cognitive modules.

If this approach is correct, it has important implications for the study of domain specificity in human cognition. In particular, it evaporates, I believe, the cultural diversity argument against the modularity of thought. For even if thought were wholly modular, we should nevertheless find many cultural domains, varying from culture to culture, whose contents are such that it would be preposterous to assume that they are the proper domain of an evolved module. The cultural idiosyncrasy and lack of relevance to biological fitness of a cognitive domain leave entirely open the possibility that it might be a domain of a genetically specified module: its cultural domain.

Meta-representational Abilities and Cultural Explosion

If you are still not satisfied that human thought could be modular through and through, if you feel that there is more integration taking place than I have allowed for so far, if you can think of domains of thought that don't fit with any plausible module, then we agree. It is not just that beliefs about Camembert cheese might play a role in forming conclusions about quarks; it is that we have no trouble at all entertaining and understanding a conceptual representation in which Camembert cheese and quarks occur simultaneously. You have just proved the point by understanding the previous sentence.

Anyhow, with or without Camembert cheeses, beliefs about quarks are hard to fit into a modular picture. Surely, they don't belong to the actual domain of naïve physics. Similarly, beliefs about chromosomes don't belong to the actual domain of naïve biology; beliefs about lycanthropy don't belong to the actual domain of folk zoology; and beliefs about the Holy Trinity or about cellular automata seem wholly removed from any module.

Is this to say that there is a whole range of extra-modular beliefs, of which many religious and scientific beliefs would be prime examples? Not really. We have not yet exhausted the resources of the modular approach. Humans have the ability to form mental representations of mental representations; in other words, they have a meta-representational ability (see chapters 3 and 4). This ability is so particular, in terms of both its domain and of its computational requirements, that anyone willing to contemplate the modularity of thought thesis will be willing to see it as modular. Even Fodor does (1992). The meta-representational module[45] is a special conceptual module, however; a second-order one, so to speak. Whereas other conceptual modules process concepts and representations of things, typically of things perceived, the meta-representational module processes concepts of concepts and representations of representations.

The actual domain of the meta-representational module is clear enough: it is the set of all representations of which the organism is capable of inferring or otherwise apprehending the existence and content. But what could be the proper domain of that module? Much current work (e.g. Astington et al. 1989) assumes that the function of the ability to form and process meta-representations is to provide humans with a naïve psychology. In other words, the module is a 'theory of mind module' (Leslie 1994), and its proper domain is that of the beliefs, desires and intentions that cause human behaviour. This is indeed highly plausible. The ability to understand and to categorize behaviour, not as mere bodily movements, but in terms of underlying mental states, is an essential adaptation for organisms that must co-operate and compete with one another in a great variety of ways.

Once you have mental states in your ontology, and the ability to attribute mental states to others, it is but a short step, or no step at all, to your having desires about these mental states – desiring that she should believe this, desiring that he should desire that – and to forming intentions to alter the mental states of others. Human communication is both a way to satisfy such meta-representational desires and an exploitation of the meta-representational abilities of one's audience. As suggested by Grice (1957) and developed by Deirdre Wilson and myself (Sperber and Wilson 1986), a communicator, by means of her communicative behaviour, is deliberately and overtly helping her addressee to infer the content of the mental representation she wants him to adopt.

Communication is, of course, radically facilitated by the emergence of a public language. A public language is rooted in another module, the language faculty. We claim, however, that the very development of a public language is not the cause, but an effect, of the development of communication made possible by the meta-representational module.

As a result of the development of communication, particularly of linguistic communication, the actual domain of the meta-representational module is teeming with representations made manifest by communicative behaviours: intentions of communicators and contents communicated. Most representations about which there is some interesting epidemiological story to be told are communicated in this manner, and therefore enter people's minds via the meta-representational module.

As already suggested, communicated contents, although they enter via the meta-representational module, may find their way to the relevant modules. What you are told about cats is integrated with what you see of cats, in virtue of the fact that the representation communicated contains the concept CAT. But now you have the information in two modes: as a representation of cats, handled by a first-order conceptual module, and as a representation of a representation of cats, handled by the second-order meta-representational module. The latter module knows nothing about cats, but it may know something about semantic relationships among representations; it may have some ability to evaluate the validity of an inference, the evidential value of some information, the relative plausibility of two contradictory beliefs, and so forth. It may also evaluate a belief, not on the basis of its content, but on the basis of the reliability of its source. The meta-representational module may therefore form or accept beliefs about cats for reasons that have nothing to do with the kind of intuitive knowledge that the CAT module (or whatever first-order module handles cats) delivers.

An organism endowed with perceptual and first-order conceptual modules has beliefs delivered by these modules, but has no beliefs about beliefs, either its own or those of others, and no reflexive attitude to them. The vocabulary of its beliefs is limited to the output vocabulary of its modules, and it cannot conceive or adopt a new concept or criticize or reject an old one. By contrast, an organism also endowed with a meta-representational module can represent concepts and beliefs *qua* concepts and beliefs, evaluate them critically, and accept them or reject them on meta-representational grounds. It may form representations of concepts and of beliefs pertaining to any conceptual domain, of a kind that the modules specialized in those domains might be unable to form on their own, or even to incorporate. In doing so, however, the better-endowed organism is merely using its meta-representational module within the module's own domain – that is, representations.

Humans, with their outstanding meta-representational abilities, may thus have beliefs pertaining to the same conceptual domain rooted in two quite different modules: the first-order module specialized in that conceptual domain and the second-order meta-representational module specialized in representations. These are, however, two different kinds of beliefs: 'intuitive beliefs' rooted in first-order modules and 'reflective beliefs' rooted in the meta-representational module (see Sperber 1985a, 1985b: ch. 2, 1990a). Reflective beliefs may contain concepts (e.g. 'quarks', 'Trinity') that do not belong in the repertoire of any module, and that are therefore available to humans only reflectively, via the beliefs or theories in which they are embedded. The beliefs and concepts that vary most from culture to culture (and that often seem unintelligible or irrational from another culture's perspective) are typically reflective beliefs and the concepts they introduce.

Reflective beliefs can be counter-intuitive – or, more precisely, they can be counter-intuitive with respect to our intuitions about their subject-matter – while, at the same time, our meta-representational reasons for accepting them are intuitively compelling. This is relevant to the most interesting of Fodor's technical arguments against the modularity of central processes. The informational encapsulation and mandatory character of perceptual modules is evidenced, Fodor points out, by the persistence of perceptual illusions, even when we are apprised of their illusory character. There is, he argues, nothing equivalent at the conceptual level. True, perceptual illusions have the feel, the vividness, of perceptual experiences that you won't find at the conceptual level. But what you do find is that you may give up a belief and still feel its intuitive force, and feel also the counterintuitive character of the belief you adopt in its stead.

You may believe with total faith in the Holy Trinity, and yet be aware of the intuitive force of the idea that a father and son cannot be one. You may understand why black holes cannot be seen, yet feel the intuitive force of the idea that a big solid — indeed, dense — object cannot but be visible. The case of naive versus modern physics provides many other blatant examples.[46] What happens, I suggest, is that the naïve physics module remains largely unpenetrated by the ideas of modern physics, and keeps delivering the same intuitions, even when they are not believed any more (or, at least, not reflectively believed).

More generally, the recognition of the meta-representational module, of the duality of beliefs that it makes possible, and of the gateway it provides for cultural contagion, plugs a major gap in the modular picture of mind that I have been trying to outline. The mind is here pictured as involving three tiers: a single, thick layer of input modules, just as Fodor says; a complex network of first-order conceptual modules of all kinds; and a second-order meta-representational module. Initially, this meta-representational module is not very different from other conceptual modules, but it allows the development of communication, and triggers a cultural explosion of such magnitude that its actual domain is blown up, and ends up hosting a multitude of cultural representations belonging to several cultural domains.

This is how you can have a truly modular mind playing a major causal role in the generation of true cultural diversity.

Conclusion: What is at Stake?

When one is attempting to throw some new light on an old domain; when one is groping in the dark, looking for marks, for paths, for bridges, no conclusion can be drawn. At each step, though, questions arise: what is to be gained? what are the risks?

What one hopes to gain is plain. One is striving for new and powerful means for understanding and explaining social phenomena. An epidemiology of representations, I insist, is intended not to replace existing means of comprehension, but to complement them. However, it would be naïve to assume that all conceivable research programmes could coexist in harmony. To begin with, in academic institutions, human and material resources are limited, and competition is unavoidable. In such competition, every party tends to push its own programme by advertising mere hopes as promises, or even as achievements, and by disparaging competing approaches.

There are also, more interestingly, genuine theoretical conflicts. For instance, Freudians and Jungians cannot simultaneously be right; nor can classical functionalists and their Marxist critics. There is no theoretical conflict — or there should not be — between interpretive approaches which aim at making social phenomena intuitively intelligible and an epidemiological approach, which seeks causal explanations. On the other hand, the epidemiological approach is, objectively, in conflict with theoretical programmes that attempt to explain social phenomena causally without reconceptualizing the domain as a whole. The specific epidemiological approach presented in this book is also to be contrasted with other naturalistic programmes that draw almost exclusively on biology, and grant only a minor, trivial role to human psychology in the explanation of cultures.

Besides issues having to do with the practice or the contents of science, we should reflect also on our very reasons for approaching social phenomena from a scientific and, more specifically, from a naturalistic point of view. Any attempt to analyse social and cultural phenomena in a scientific manner, in particular any naturalistic attempt, is sure to meet with accusations of reductionism. Of course, such accusations could be brushed aside. It is not hard to show that the label 'reductionist' is doubly misused: on the one hand, nobody is really proposing a reduction of social phenomena; on the other, should such a proposition be made seriously, it would deserve interest rather than scorn, since true reductions are major scientific advances. One might also argue that a naturalistic analysis of mental and social mechanisms, far from belittling them, as is sometimes feared, would tend, rather, to highlight their wealth and their subtlety. One might challenge the critics: let them show that the so-called reductionist approaches are ill-conceived, or else let them articulate the moral reasons for their censorship.

Still, even if they are poorly articulated, even if they don't provide any serious arguments against the pursuit of the project, these accusations of reductionism spring, I believe, from a legitimate uneasiness. Any research involves responsibilities and risks. In the social sciences, responsibilities and risks are moral and political. Modern political movements, be they reactionary, conservative, progressive, or revolutionary, draw on theories in the social sciences and claim the 'scientific' character of these theories as a source of legitimacy. Often social scientists themselves encourage such a use of their theories. That science should guide action, what could be more desirable? However, misuses of science are frequent; they range from arrogance to crime. It is reasonable, therefore, to be on guard. A naturalistic approach to the social gives rise to misgivings of two kinds, some having to do with the role given to biology, others with what may look like scientism.

The social sciences, and in particular anthropology, have had their share of responsibility in the crimes of colonialism and racism. Typically, exploitation and extermination have been justified in the name of an alleged biological superiority of exploiters over the exploited, and of exterminators over their victims. There are, still today, people who, in the name of biology, advocate various forms of social or racial discrimination. Does this imply that any appeal to biology in the human sciences cannot but open the door to racism? Such a suspicion should itself be suspect, for it would make sense only if the study of biology did, in fact, provide arguments in favour of racism.

Actually, there is a radical and blatant difference between the scientific goals of serious evolutionary anthropology and psychology and the pseudo-scientific concerns of people motivated by an attraction to racism. What may contribute to a better understanding of human affairs is a biological perspective on what humans have in common. We want to understand what, in the genetic make-up of the species, readies its members for social and cultural life. What racists are seeking are biological differences among human groups that would explain and justify their unequal fates. Even without going into the detail of evidence and arguments (but see Cavalli-Sforza et al. 1994), it should be manifest that the racists' quest lacks scientific merit.

Every human being (except for identical twins) is genetically different from all others. Yet all human beings are genetically similar enough to be able to learn any human language and to acquire any human culture. This is not a mere theoretical possibility; it is a fact essential to the understanding of human history. In human history, population movements from one society to another occur all the time. Such movements both presuppose and maintain the fundamental unity of the species. Because of this, humankind is not divided in distinct, homogeneous genetic subgroups; there are no human races. Whatever genetic differences may be found between human groups are shallow and transitory. Such differences could not play more than an utterly marginal role in the (yet to be given) explanation of the diversity of human cultures.

Those who nevertheless choose to invest their energy in the search for a genetic explanation of cultural and historical differences between groups, in spite of the lack of any scientific justification for such an investment, are either inept researchers, or, more probably, racists trying to make their cause benefit from the prestige and authority of science, without submitting to its requirements of objectivity and fecundity. Such pseudo-scientists should be kept at bay. This can be done simply by maintaining high scientific standards, especially so when would-be scientific claims have significant social implications (just as, in all sciences, research likely to affect human interests is evaluated with particular strictness).

A naturalistic approach may cause other misgivings, more diffuse maybe, but nevertheless pertinent. We all have, as social actors, some understanding of the mechanisms of social life, which helps us evaluate choices, decide and act. In a democratic society, choices are not just about social relationships in which we are involved personally; they are also about the becoming of the society at large. Our individual practice gives us only a rudimentary, and probably biased, understanding of global social phenomena. The social sciences, therefore, have a fundamental role to play in democratic life: that of enlightening citizens.

In a large part, research in the social sciences is done at the request of political actors: citizens, militants, and authorities. There is a welcome continuity between much of this research and our common-sense understanding of the social world. Truly theoretical research in the social sciences does not answer the same kind of social need. Nevertheless, because of the concepts it uses are commonsensical, much theoretical research is intelligible to lay readers. One may wonder whether this would be true also of a naturalistic programme such as the one I am advocating. After all, it calls for a recasting of the very concepts ordinarily used to think about social matters.

Should one take the chance of pulling the social sciences away from common-sense understanding? Even if naturalistic programmes are, at best, in their infancy, isn't there a risk that one day scientists who pursue such programmes may present themselves as experts who argue among experts and presume to decide for the rest of us? The danger may be remote, but still, such worries are not absurd. As a result, some might draw the conclusion that any naturalistic programme in the social sciences should be opposed from the start. I myself conclude that a plurality of methods and points of views must be vigorously defended. Pluralism is essential in the sciences generally, since it is a condition of their progress. Every novel and potentially fruitful perspective is worth exploring. Pluralism is doubly essential in the social sciences, who must – not individually, but jointly – answer different social demands, and in particular the demand for intelligibility fostered by democracy, however imperfect.

A naturalistic programme may evoke yet another concern. As in a mirror, we look for our image in the social sciences. When we do not recognize ourselves in the reflected image, we are disturbed. Cognitive psychology does not reflect an immediately recognizable image of ourselves; nor would an epidemiology of representations. Worse, what we think of as essential and primary – that is, our existence as conscious persons – comes out, at best, as a changing pattern, socially projected on to a biological structure, itself precarious. Should we ever have to content ourselves with this single, disquieting picture, there might be cause for alarm. However, the more common images we have of ourselves are under no threat. Modern physics leaves essentially untouched the image of the material world whereby we guide our steps. Similarly, no future social science will displace our common-sense understanding of ourselves. At most, science will put common sense in perspective.

The sciences are capable of giving us a special kind of intellectual pleasure: that of seeing the world in a light that at first disconcerts, but then forces reflection, and deepens our knowledge while relativizing it. I wish the social sciences would, more often, give us pleasure of this kind.

Notes

1 A more exact term would be 'physicalist' (one who believes that everything that exists exists physically, leaving it to physicists to explain what 'physically' means), since the very notion of matter implied in 'materialist' is an unclear one. However, the old term 'materialist' is better known, especially in the social sciences, and, anyhow, subtle considerations on the place of matter in physics are irrelevant to my present purpose.

2 You might, for instance, adopt Searle's (1969) approach to 'institutional facts' and, roughly, define 'being married' as being believed by the right people to be married. This definition, of course, is not the native's. It still has all the fuzziness problems of family resemblance notions. It has problems of circularity of its own. Nor is it being used in any insightful theory.

3 On the distinction between interpretation and description see Sperber 1985b, ch. 1, and Sperber and Wilson 1986: ch. 4.

4 For a discussion of Dumézil's approach and a comparison with Lévi-Strauss's, see Smith and Sperber 1971.

5 The weaknesses of functionalist typologies have been discussed by Leach 1961 and, more thoroughly, by Needham 1971, 1972. I have argued that these unprincipled, fuzzy typologies are based on interpretive rather than descriptive criteria; see ch. 1.

6 The relevant facts have been highlighted in the work of Daniel Kahneman and Amos Tversky (see Kahneman et al. 1982), Gerd Gigerenzer and his collaborators, and in the debate between the two approaches (see Gigerenzer, 1991, 1993; Gigerenzer and Hoffrage 1995; Kahneman and Tversky, forthcoming).

7 Some ethnographic studies have stressed the micro-mechanisms of cultural transmission and are of particular interest for the epidemiological approach. I will mention just two classics: Barth 1975; Favret-Saada 1980.

8 See e.g. Levine 1973; Jahoda 1982.

9 I do not mean to imply that the psychology of emotions is irrelevant to the explanation of culture. I tend to believe, though, that important advances are needed on the cognitive side in order better to understand the role of emotion

in culture. For recent discussions see Lewis 1977; Schweder 1979a, 1979b, 1980; D'Andrade 1981; Gibbard 1990.

10 For an introduction to epidemiology see MacMahon and Pugh 1970.

11 As exemplified by recent work in the philosophy of biology; see Darden and Maull 1977; Darden 1978.

12 For two different versions of the Platonist approach see Popper 1972 and Katz 1981.

13 See Tyler 1969.

14 E.g. Vygotsky 1965; Bruner et al. 1956.

15 See e.g. Berlin and Kay 1969; Miller and Johnson-Laird 1976; Rosch and Lloyd 1978; Keil 1979; Ellen and Reason 1979; Smith and Medin 1981; and the recent synthetical papers of Scott Atran (1981, 1983, 1987).

16 E.g. Bloch 1977; Sperber 1975b; 1985b.

17 In particular Colby and Cole 1973. Lévi-Strauss (esp. 1971) has alluded to the role of memory in shaping myths, but without going into the psychology of memory at all. See Sperber 1975b, 1985b: ch. 3 for a discussion of his contribution.

18 See e.g. Rumelhardt 1975; Kintsch 1977; Mandler and Johnson 1977; van Dijk 1980; Brewer and Lichtenstein 1981; Wilensky 1983.

19 See Goody 1977 for an anthropological discussion.

20 The parallelism is overstated: your mental representations do not represent something *for* you in the same way that these words represent something *for* you, but nothing essential to the present discussion hinges on that.

21 The philosophical literature on beliefs is huge (see e.g. Ryle 1949; Hintikka 1962; Armstrong 1973; Harman 1973; Dennett 1978; Dretske 1981; Stich 1983; Bogdan 1986; Brandt and Harnish 1986). However, it pays little or no attention to those features of beliefs which social scientists have particularly been concerned with. Though 'belief' has always been a stock-in-trade term of anthropologists, Needham 1972 is the only thorough discussion of the concept from an anthropological point of view (inspired by Wittgenstein).

22 'Box', of course, is to be loosely understood: rather than correspond to a location in the brain, it might refer, say, to a way of indexing representations. So understood, the belief box story is not terribly novel or controversial, but it helps bring into focus what is generally merely presupposed.

23 On this contrast between description and interpretation, see Sperber 1985b, ch. 1 and Sperber and Wilson 1986: ch. 4.

24 In Sperber 1975b, I describe reflective beliefs as being 'in quotation marks'; in Sperber 1985b, I contrast 'factual' and 'representational' beliefs; in Sperber 1985a, reproduced here, with minor changes, as chapter 3, I contrast 'basic' and 'speculative' beliefs. Each of these terminologies turned out to be misleading in some ways. I hope the present proposal won't be.

25 There are grounds for considering that intuitive beliefs in different cognitive domains – naïve physics, naïve zoology, naïve psychology – have different conceptual structures (see Sperber 1975a, Atran 1987, Atran and Sperber 1991).

These differences, however, do not seem to be such as to bring about very different modes of distribution.

26 Work on very early cognitive development (e.g. Spelke 1988), shows that infants have definite expectations about, e.g., the movement of objects which they could not possibly owe to communication. If these expectations are intuitive beliefs in the relevant sense, and are not, later on, superseded by language-influenced intuitive beliefs, then some intuitive beliefs of adult humans are quite independent of communication.

27 Of course, it is contentious that mental things can be naturalized. If they cannot, if no bridges can be constructed between the psychological and the neurological levels, then this reduces to a proposal to bridge the social with the psychological and the ecological sciences. If, as I believe, a naturalistic programme in psychology is well under way, then this is a proposal to bridge the social and the natural sciences through psychology and ecology.

28 See Sober 1991.

29 On the other hand, the view of cultural evolution put forward by Pascal Boyer (1993: ch. 9) and my own are very close. Boyer's arguments and mine are in part similar, in part complementary. Boyer offers a detailed discussion of the models of Lumsden and Wilson (1981), Boyd and Richerson (1985), and Durham (1991). There is also a good deal of convergence with Tooby and Cosmides (1992). Two other original and important approaches, that of the cognitive anthropologist Ed Hutchins (1994), and that of the philosopher Ruth Millikan (1984, 1993) would deserve a separate discussion.

30 This is how I seem to be interpreted by Dennett (1995: 357–9).

31 A theme developed in detail by Millikan (1984).

32 Williams goes as far as to suggest that, 'In evolutionary theory, a gene could be defined as any hereditary information for which there is a favourable or unfavourable selection bias equal to several or many times its rate of endogenous change. The prevalence of such stable entities in the heredity of populations is a measure of the importance of natural selection' (1966: 25).

33 See Wilson and Bossert 1971: 61–2; Maynard Smith 1989: 20–4.

34 Sophisticated notions of attractors ('strange attractors' in particular) have been developed in complex systems dynamics, and may well turn out to be of future use in modelling cultural evolution, but a very elementary notion of an attractor will do for the present purpose.

35 For this chapter, I have benefited from useful comments by Ned Block, John Maynard Smith, and Elliott Sober.

36 Fodor also mentions the possibility that output – motor – systems might be modular too. I assume that this is so, but will not discuss the issue here.

37 See also Rozin 1976; Symons 1979; Rozin and Schull 1988; Barkow 1989; Brown 1991; Barkow et al. 1992.

38 The point cannot just be that the forces that have driven cognitive evolution cannot be identified with certainty; that much is trivially true. The claim must be that these forces cannot be even tentatively and reasonably identified, unlike

the forces that have driven the evolution of, say, organs of locomotion. See Piatelli-Palmarini 1989 and Stich 1990 for clever, but unconvincing, arguments in favour of this second law.

39 Putnam 1975 and Burge 1979 offered the initial arguments for externalism (I myself am convinced by Putnam's but not by Burge's). For a sophisticated discussion, see Recanati 1993.

40 Arguably, content is a biological function in an extended sense – see Dennett 1987; Dretske 1988; Millikan 1984; Papineau 1987. My views have been influenced by Millikan's.

41 There are, of course, conceptual problems here (see Dennett 1987; Fodor 1988). It could be argued, for instance, that the module's proper domain was neither elephants nor hippos, but something else, say, 'approaching big animals that might trample orgs'. If so, we would want to say that its proper domain had *not* changed with the passing of the elephants and the coming of the hippos. I side with Dennett in doubting that much of substance hinges on which of these descriptions we choose: the overall explanation remains exactly the same.

42 This is why it would be a mistake to say that the function of a device is to react to whatever might satisfy its input conditions, and to equate its actual and proper domains. Though there may be doubt about the correct assignment of the proper domain of some device (see the preceding note), the distinction between actual and proper domains is as solid as that between effect and function.

43 Here, as in talk of representations competing for attention, the term 'competition' is only a vivid metaphor: no intention or disposition to compete is implied. What is meant is that, out of all the representations present in a human group at a given time, some, at one extreme, will spread and last, while others, at the opposite extreme, will occur only very briefly and very locally. This is not a random process, and it is assumed that properties of the information play a causal role in determining its wide or narrow distribution.

44 See Pinker and Bloom 1990 and my contribution to the discussion of their paper (Sperber 1990b).

45 The capacity to form and process meta-representations could be instantiated not in a single, but in several distinct modules, each, say, meta-representing a different domain or type of representations. For lack of compelling arguments, I will ignore this genuine possibility.

46 And a wealth of subtler examples have been analysed in a proper cognitive perspective by Atran (1990).

References

Armstrong, D. (1973). *Belief, Truth and Knowledge*. Cambridge: Cambridge University Press.

Astington, J.W., Harris, P. and Olson, D. (1989). *Developing Theories of Mind*. Cambridge: Cambridge University Press.

Atran, S. (1981). Natural Classification. *Social Science Information*, 20, 27–91.

Atran, S. (1983). Covert Fragments and the Origins of the Botanical Family. *Man*, n.s. 18, 51–71.

Atran, S. (1985). The Nature of Folk-botanical Life-forms. *American Anthropologist*, 87, 298–315.

Atran, S. (1986). *Fondements de l'histoire naturelle*. Brussels: Complexe.

Atran, S. (1987). Constraints on the Ordinary Semantics of Living Kinds. *Mind and Language*, 2(1), 27–63.

Atran, S. (1990). *Cognitive Foundations of Natural History*. Cambridge: Cambridge University Press.

Atran, S. (1994). Core Domains versus Scientific Theories. In L. A. Hirschfeld and S. A. Gelman (eds), *Mapping the Mind: Domain Specificity in Cognition and Culture*. New York: Cambridge University Press, 316–40.

Atran, S. and Sperber, D. (1991). Learning without Teaching: Its place in Culture, In L. Landsmann (ed.), *Culture, Schooling and Psychological Development*. Norwood, N.J.: Ablex, 39–55.

Barkow, J. H. (1989). *Darwin, Sex and Status: Biological Approaches to Mind and Culture*. Toronto: University of Toronto Press.

Barkow, J. H., Cosmides, L. and Tooby, J. (eds) (1992). *The Adapted Mind: Evolutionary Psychology and the Generation of Culture*. New York: Oxford University Press.

Barth, F. (1975). *Ritual and Knowledge among the Baktaman of New Guinea*. New Haven, Conn.: Yale University Press.

Berlin, B. (1978). Ethnobiological Classification. In E. Rosch and B. Lloyd (eds), *Cognition and Categorization*. Hillsdale, N.J.: Lawrence Erlbaum Associates, 9–26.

Berlin, B. and Kay, P. (1969). *Basic Color Terms*. Berkeley: University of California Press.

Berlin, B., Breedlove, D. and Raven, P. (1973). General Principles of Classification and Nomenclature in Folk Biology. *American Anthropologist*, 75, 214–42.

Bloch, M. (1977). The Past and the Present in the Present. *Man*, n.s. 12, 278–92.

Bloch, M. (1983). *Marxism and Anthropology*. Oxford: Oxford University Press.

Block, N. (1980). *Readings in Philosophy of Psychology*, vol I. Cambridge, Mass.: Harvard University Press.

Bogdan, R. (ed.) (1986). *Belief: Form, Content and Function*. Oxford: Oxford University Press.

Boyd, R. and Richerson, P. J. (1985). *Culture and the Evolutionary Process*. Chicago: University of Chicago Press.

Boyer, P. (1990). *Tradition as Truth and Communication*. Cambridge: Cambridge University Press.

Boyer, P. (1994). *The Naturalness of Religious Ideas*. Berkeley: University of California Press.

Boyer, P. (1994). Cognitive Constraints on Cultural Representations: Natural Ontologies and Religious Ideas. In L. A. Hirschfeld and S. A. Gelman (eds), *Mapping the Mind: Domain Specificity in Cognition and Culture*. New York: Cambridge University Press, 391–411.

Brandt, M. and Harnish, R. M. (eds) (1986). *The Representation of Knowledge and Belief*. Tucson: University of Arizona Press.

Brewer, W. F. and Lichtenstein, E. H. (1981). Event Schemas, Story Schemas, and Story Grammars. In J. Long and A. Baddeley (eds), *Attention and Performance*, vol. 9. Hillsdale, N.J.: Lawrence Erlbaum Associates, 363–79.

Brown, A. (1991). *Human Universals*. New York: McGraw-Hill.

Bruner, J. S., Goodnow, J. J. and Austin, G. A. (1956). *A Study of Thinking*. New York: Wiley.

Burge, T. (1979). Individualism and the Mental. *Midwest Studies in Philosophy*, 5, 73–122.

Campbell, D. T. (1974). Evolutionary Epistemology. In P. A. Schilpp (ed.), *The Philosophy of Karl Popper*. La Salle, Ill.: Open Court, 413–63.

Carey, S. (1982). Semantic Development: The State of the Art. In E. Wanner and L. Gleitman (eds), *Language Acquisition: The State of the Art*. Cambridge: Cambridge University Press, 347–89.

Carey, S. (1985). *Conceptual Change in Childhood*. Cambridge, Mass.: MIT Press.

Cavalli-Sforza, L. L. and Feldman, M. W. (1981). *Cultural Transmission and Evolution: A Quantitative Approach*. Princeton: Princeton University Press.

Cavalli-Sforza, L. L., Menozzi, P. and Piazza, A. (1994). *The History and Geography of Human Genes*. Princeton: Princeton University Press.

Chomsky, N. (1972). *Language and Mind*. New York: Harcourt Brace Jovanovich.

Chomsky, N. (1975). *Reflections on Language*. New York: Pantheon.

Chomsky, N. (1986). *Knowledge of Language: Its Nature, Origin, and Use*. New York: Praeger.

Churchland, P. M. (1988). *Matter and Consciousness*, 2nd edn. Cambridge, Mass.: MIT Press.

Clark, A. (1987). The Kludge in the Machine. *Mind and Language*, 2 (4), 277–300.

Clark, A. (1990). *Microcognition: Philosophy, Cognitive Science, and Parallel Distributed Processing*. Cambridge, Mass.: MIT Press.

Colby, B. and Cole, M. (1973). Culture, Memory, and Narrative. In R. Horton and R. Finnegan (eds), *Modes of Thought*. London: Faber, 63–91.

Cosmides, L. (1989). The Logic of Social Exchange: Has Natural Selection Shaped how Humans Reason? Studies with the Wason Selection Task. *Cognition*, 31, 187–276.

Cosmides, L. and Tooby, J. (1987). From Evolution to Behavior: Evolutionary Psychology as the Missing Link. In J. Dupré (ed.), *The Latest on the Best: Essays on Evolution and Optimality*. Cambridge Mass.: MIT Press, 277–306.

Cosmides, L. and Tooby, J. (1994). Origins of Domain-Specificity: The Evolution of Functional Organization. In L. A. Hirschfeld and S. A. Gelman (eds), *Mapping the Mind: Domain Specificity in Cognition and Culture*. New York: Cambridge University Press, 85–116.

D'Andrade, R. G. (1981). The Cultural Part of Cognition. *Cognitive Science*, 5, 179–95.

Darden, L. (1978). Discoveries and the Emergence of New Fields in Science. In P. Asquith and I. Hacking (eds), *PSA, 1978*, vol. 1. East Lansing, Mich.: Philosophy of Science Association, 149–60.

Darden, L. and Maull, N. (1977). Interfield Theories. *Philosophy of Science*, 44, 43–64.

Davidson, N. S. and Gelman, S. (1990). Induction from Novel Categories: The Role of Language and Conceptual Structure. *Cognitive Development*, 5, 121–52.

Dawkins, R. (1976). *The Selfish Gene*. Oxford: Oxford University Press.

Dawkins, R. (1982). *The Extended Phenotype*. Oxford: Oxford University Press.

Dehaene, S. (1992). Varieties of Numerical Abilities. *Cognition*, 44 (1–2), 1–42.

Dennett, D. (1978). *Brainstorms*. Hassocks: Harvester Press.

Dennett, D. (1987). *The Intentional Stance*. Cambridge, Mass.: MIT Press.

Dennett, D. (1991). *Consciousness Explained*. Boston: Little, Brown.

Dennett, D. (1995). *Darwin's Dangerous Idea: Evolution and the Meaning of Life*. New York: Simon and Schuster.

Détienne, M. (1981). *L'Invention de la mythologie*. Paris: Gallimard.

Dijk, T. A. van (1980). Story Comprehension: An Introduction. *Poetics*, 9, 1–21.

Douglas, M. (1966). *Purity and Danger: An Analysis of the Concepts of Pollution and Taboo*. London: Routledge and Kegan Paul.

Douglas, M. (1975). *Implicit Meanings*. London: Routledge and Kegan Paul.

Dretske, F. (1981). *Knowledge and the Flow of Information*. Cambridge, Mass.: MIT Press.

Dretske, F. (1988). *Explaining Behavior*. Cambridge, Mass.: MIT Press.

Dumézil, G. (1968). *Mythe et épopée: L'idéologie des trois fonctions dans les épopées des peuples Indo-Europeens*. Paris: Gallimard.

Durham, W. H. (1991). *Coevolution: Genes, Culture and Human Diversity*. Stanford, Calif.: Stanford University Press.

Ellen, R. and Reason, D. (eds) (1979). *Classifications in their Social Context*. London: Academic Press.

Favret-Saada, J. (1980). *Deadly Words*. Cambridge: Cambridge University Press.

Fodor, J. (1974). Special Sciences. *Synthese*, 28, 77–115. Repr. in Fodor 1981, 127–45.

Fodor, J. (1975). *The Language of Thought*. New York: Crowell.

Fodor, J. (1981). *Representations: Philosophical Essays on the Foundations of Cognitive Science*. Cambridge, Mass.: MIT Press.

Fodor, J. (1983). *The Modularity of Mind*. Cambridge, Mass.: MIT Press.

Fodor, Jerry (1987a). Modules, Frames, Fridgeons, Sleeping Dogs, and the Music of the Spheres. In J. Garfield (ed.), *Modularity in Knowledge Representation and Natural-Language Understanding*. Cambridge Mass.: MIT Press, 26–36.

Fodor, J. (1987b). *Psychosemantics*. Cambridge, Mass.: MIT Press.

Fodor, J. (1988). *Psychosemantics*. Cambridge, Mass.: MIT Press.

Fodor, J. (1992). A Theory of the Child's Theory of Mind. *Cognition*, 44, 283–96.

Gallistel, C. R. and Gelman, R. (1992). Preverbal and Verbal Counting and Computation. *Cognition*, 44 (1–2), 43–74.

Gamst, F. and Norbeck, E. (1976). *Ideas of Cultures*. New York: Holt, Rinehart and Winston.

Geertz, C. (1973). *The Interpretation of Cultures*. New York: Basic Books.

Gelman, R. and Gallistel C. R. (1978). *The Child's Understanding of Number*. Cambridge, Mass.: Harvard University Press.

Gelman R. and Spelke, E. (1981). The Development of Thoughts about Animate and Inanimate Objects. In J. Flavell and L. Ross (eds), *Social Cognitive Development*. Cambridge: Cambridge University Press, 43–66.

Gelman, S and Coley, J. D. (1991). The Acquisition of Natural Kind Terms. In S. Gelman and J. Byrnes (eds), *Perspectives on Language and Thought*. New York: Cambridge University Press, 146–96.

Gelman, S. and Markman, E. (1986). Categories and Induction in Young Children. *Cognition*, 23, 183–209.

Gelman, S. and Markman, E. (1987). Young Children's Inductions from Natural Kinds: The Role of Categories and Appearances. *Child Development*, 58, 1532–41.

Gibbard, A. (1990). *Wise Choices, Apt Feelings*. Cambridge, Mass.: Harvard University Press.

Gigerenzer, G. (1991). How to Make Cognitive Illusions Disappear: Beyond 'Heuristics and biases'. In W. Stroche and M. Hewstone (eds), *European Review of Social Psychology*, vol. 2. Chichester: Wiley, 83–115.

Gigerenzer, G. (1993). The Bounded Rationality of Probabilistic Mental Models. In K. I. Manktelow and D. E. Over (eds), *Rationality: Psychological and Philosophical Perspectives*. London: Routledge, 284–313.

Gigerenzer, G. and Hoffrage, U. (1995). How to Improve Bayesian Reasoning without Instructions: Frequency Formats. *Psychological Review*, 102 (4), 684–704.

Goldenweiser, A. (1910). Totemism: An Analytical Study. *Journal of American Folklore*, 23, 178–298.

Goody, J. (1977). *The Domestication of the Savage Mind*. Cambridge: Cambridge University Press.

Grice, H. P. (1957). Meaning. *Philosophical Review*, 66, 377–88.

Harman, G. (1973). *Thought*. Princeton, N.J.: Princeton University Press.

Hintikka, J. (1962). *Knowledge and Belief*. Ithaca, N.Y.: Cornell University Press.

Hirschfeld, L. (1984). Kinship and Cognition. *Current Anthropology*, 27(3), 217–42.

Hirschfeld, L. (1988). On Acquiring Social Categories: Cognitive Development and Anthropological Wisdom. *Man*, n.s. 23, 611–38.

Hirschfeld, L. (1993). Discovering Social Difference: The Role of Appearance in the Development of Racial Awareness. *Cognitive Psychology*, 25, 317–50.

Hirschfeld, L. (1994). The Acquisition of Social Categories. In L. A. Hirschfeld and S. A. Gelman (eds), *Mapping the Mind: Domain Specificity in Cognition and Culture*. New York: Cambridge University Press, 201–33.

Hirschfeld, L. A. and Gelman, S. A. (eds) (1994). *Mapping the Mind: Domain Specificity in Cognition and Culture*. New York: Cambridge University Press.

Hutchins, E. (1994). *Cognition in the Wild*. Cambridge, Mass.: MIT Press.

Jahoda, G. (1982). *Psychology and Anthropology*. London: Academic Press.

Kahneman, D., Slovic, P. and Tversky, A. (1982). *Judgement under Uncertainty: Heuristics and Biases*. Cambridge: Cambridge University Press.

Kahneman, D. and Tversky, A. (in press). On the Reality of Cognitive Illusions: A Reply to Gigerenzer's Critique. *Psychological Review*.

Kaplan, D. (1965). The Superorganic: Science or Metaphysics? *American Anthropologist*, 67(4), 958–76.

Katz, J. J. (1981). *Language and Other Abstract Objects*. Totowa, N.J.: Rowman and Littlefield.

Keil, F. C. (1979). *Semantic and Conceptual Development*. Cambridge, Mass.: Harvard University Press.

Keil, F. C. (1989). *Concepts, Kinds, and Cognitive Development*. Cambridge, Mass.: Bradford Book/MIT Press.

Keil, F. C. (1994). The Birth and Nurturance of Concepts by Domains: The Origins of Concepts of Living Things. In L. A. Hirschfeld and S. A. Gelman (eds), *Mapping the Mind: Domain Specificity in Cognition and Culture*. New York: Cambridge University Press, 234–54.

Kelly, M. and Keil, F. C. (1985). The More Things Change . . . : Metamorphoses and Conceptual Development. *Cognitive Science*, 9, 403–16.

Kintsch, W. (1977). Understanding Stories. In M. Just and P. Carpenter (eds), *Cognitive Processes in Comprehension*. Hillsdale, N.J.: Lawrence Erlbaum Associates, 33–62.

Kroeber, A. L. and Kluckhohn, C. (1952). *Culture: A Critical Review of Concepts and Definitions*. Papers of the Peabody Museum, 47(1), 1–223.

Leach, E. (1961). *Rethinking Anthropology*. London: Athlone Press.

Leslie, A. (1987). Pretense and Representation: The Origins of 'Theory of Mind'.

Psychological Review, 94, 412–26.

Leslie, A. (1988). The Necessity of Illusion: Perception and Thought in Infancy. In L. Weiskrantz (ed.), *Thought without Language*. Oxford: Clarendon Press, 185–210.

Leslie, A. (1994). ToMM, ToBY, and Agency: Core architecture and Domain Specificity. In L. A. Hirschfeld and S. A. Gelman (eds), *Mapping the Mind: Domain Specificity in Cognition and Culture*. New York: Cambridge University Press, 119–48.

Levine R. A. (1973). *Culture, Behaviour, and Personality*. London: Hutchinson.

Lévi-Strauss, C. (1956). The Family. In H. L. Shapiro (ed.), *Man, Culture and Society*. Oxford: Oxford University Press, 261–85.

Lévi-Strauss, C. (1963). *Structural Anthropology*. New York: Basic Books.

Lévi-Strauss, C. (1966a). *The Savage Mind*. Chicago: University of Chicago Press.

Lévi-Strauss, C. (1966b). *Totemism*. Boston: Beacon Press.

Lévi-Strauss C. (1971). *L'Homme nu: Mythologiques IV*. Paris: Plon.

Lévi-Strauss, C. (1973). *Anthropologie structurale Deux*. Paris: Plon.

Levy, R. (1984). The emotions in comparative perspective. In K. R. Scherer and P. Eckman (eds), *Approaches to Emotion*. Hillsdale, N.J.: Erlbaum Associates, 397–412.

Lewis, I. M. (ed.) (1977). *Symbols and Sentiments: Cross-cultural Studies in Symbolism*. London: Academic Press .

Lumsden, C. J. and Wilson, E. O. (1981). *Genes, Mind and Culture*. Cambridge, Mass.: Harvard University Press.

MacMahon, B. and Pugh, T. F. (1970). *Epidemiology: Principles and Methods*. Boston: Little, Brown.

Mandler, J. M. and Johnson, N. S. (1977). Remembrance of Things Parsed: Story Structure and Recall. *Cognitive Psychology*, 9, 111–51.

Manktelow, K. and Over, D. (1990). *Inference and Understanding: A Philosophical and Psychological Perspective*. London: Routledge.

Markman, E. M. (1990). The Whole-Object, Taxonomic, and Mutual Exclusivity Assumptions as Initial Constraints on Word Meanings. In S. Gelman and J. Byrnes (eds), *Perspectives on Language and Thought*. New York: Cambridge University Press, 72–106.

Markman, E. M. and Hutchinson, J. E. (1984). Children's Sensitivity to Constraints on Word Meaning: Taxonomic versus Thematic Relations. *Cognitive Psychology*, 16, 1–27.

Maynard Smith, J. (1989). *Evolutionary Genetics*. Oxford: Oxford University Press.

Medin, D. and Ortony, A. (1989). Psychological Essentialism. In S. Vosniadou and A. Ortony (eds), *Similarity and Analogical Reasoning*. Cambridge: Cambridge University Press, 179–96.

Menget, P. (1982). Time of Birth, Time of Being: The Couvade. In M. Izard and P. Smith (eds), *Between Belief and Transgression: Structuralist Essays in Religion, History, and Myth*. Chicago: University of Chicago Press,193–209.

Miller, G. A. and Johnson-Laird, P. (1976). *Language and Perception*. Cambridge: Cambridge University Press.

Millikan, R. G. (1984). *Language, Thought, and other Biological Categories.* Cambridge, Mass.: MIT Press.

Millikan, R. G. (1993). *White Queen Psychology and Other Essays for Alice.* Cambridge, Mass.: MIT Press.

Nagel, E. (1961). *The Structure of Science.* New York: Harcourt Brace and World.

Needham, R. (ed.) (1971). *Rethinking Kinship and Marriage.* London: Tavistock.

Needham, R. (1972). *Belief, Language and Experience.* Oxford: Blackwell.

Needham, R. (1975). Polythetic Classification. *Man,* n.s. 10, 349–69.

Needham, R. (1981). *Circumstantial Deliveries.* Berkeley: University of California Press.

Nowak, A., Szamrej, J. and Latané, B. (1990). From Private Attitude to Public Opinion: A Dynamic Theory of Social Impact. *Psychological Review,* 97, 362–76.

Papineau, D. (1987). *Reality and Representation.* Oxford: Blackwell.

Piatelli-Palmarini, M. (1989). Evolution, Selection and Cognition: From 'Learning' to Parameter Setting in Biology and the Study of Language. *Cognition,* 31, 1–44.

Pinker, S. (1994). *The Language Instinct.* New York: Morrow.

Pinker, S. and Bloom, P. (1990). Natural Language and Natural Selection. *Behavioral and Brain Sciences,* 13(4), 756–58.

Popper K. (1972). *Objective Knowledge: An Evolutionary Approach.* Oxford: Clarendon Press.

Premack, D. (1990). The Infant's Theory of Self-propelled Objects. *Cognition,* 36, 1–16.

Premack, D. and Woodruff, G. (1978). Does the Chimpanzee have a Theory of Mind? *Behavioral and Brain Sciences,* 1(4), 515–26.

Putnam, H. (1975). The Meaning of 'Meaning.' In *Mind, Language and Reality: Philosophical Papers,* vol. 2. Cambridge: Cambridge University Press, 215–71.

Recanati, F. (1993). *Direct Reference, Meaning and Thought.* Oxford: Blackwell.

Rivière, P. (1971). Marriage: A Reassessment. In R. Needham (ed.), *Rethinking Kinship and Marriage.* London: Tavistock, 57–74.

Rivière, P. (1974). The Couvade: A Problem Reborn. *Man,* n.s. 9(3), 423–35.

Rosch, E. and Lloyd, B. (eds) (1978). *Cognition and Categorization.* Hillsdale, N.J.: Lawrence Erlbaum Associates.

Royal Anthropological Institute (1951). *Notes and Queries in Anthropology,* 6th edn, London.

Rozin, P. (1976). The Evolution of Intelligence and Access to the Cognitive Unconscious. In J. M. Sprague and A. N. Epstein (eds), *Progress in Psychobiology and Physiological Psychology.* New York: Academic Press, 245–80.

Rozin, P. and Schull, J. (1988). The Adaptive-Evolutionary Point of View in Experimental Psychology. In R. Atkinson, R. Herrnstein, G. Lindzey and R. Luce (eds), *Steven's Handbook of Experimental Psychology.* New York: John Wiley and Sons, 503–62.

Rumelhardt, D. E. (1975). Notes on a Schema for Stories. In D. G. Bobrow and A. Collins (eds), *Representation and Understanding: Studies in Cognitive Science.* New York: Academic Press, 211–36.

Ryle, G. (1949). *The Concept of Mind.* London: Hutchinson.

Schweder, R. A. (1979a). Rethinking Culture and Personality Theory, 1. *Ethos*, 7, 255–78.

Schweder, R. A. (1979b). Rethinking Culture and Personality Theory, 2. *Ethos*, 7, 279–311.

Schweder R. A. (1980). Rethinking Culture and Personality Theory, 3. *Ethos*, 8, 60–94.

Searle, J. (1969). *Speech Acts.* Cambridge: Cambridge University Press.

Smith, E. E. and Medin, D. L. (1981). *Categories and Concepts.* Cambridge, Mass.: Harvard University Press .

Smith, P. and Sperber, D. (1971). Mythologiques de Georges Dumézil. *Annales*, 26, 559–86.

Sober, E. (1984). *The Nature of Selection.* Cambridge, Mass.: MIT Press.

Sober, E. (1991). Models of Cultural Evolution. In P. Griffiths (ed.), *Trees of Life: Essays in the Philosophy of Biology.* Dordrecht: Kluwer, 17–38.

Spelke, E. S. (1988). The Origins of Physical Knowledge. In L. Weiskrantz (ed.), *Thought without Language.* Oxford: Clarendon Press, 168–84.

Sperber, D. (1968). Le Structuralisme en anthropologie. In O. Ducrot et al., *Qu'est-ce que le structuralisme?* Paris: Le Seuil, 167–238.

Sperber, D. (1974). Contre certains a priori anthropologiques. In E. Morin and M. Piatelli-Palmarini (eds), *L'Unité de l'homme.* Paris: Le Seuil, 491–512.

Sperber, D. (1975a). Pourquoi les animaux parfaits, les hybrides et les monstres sont-ils bons à penser symboliquement? *L'Homme*, 15(2), 5–24.

Sperber, D. (1975b). *Rethinking Symbolism.* Cambridge: Cambridge University Press.

Sperber, D. (1980) Is Symbolic Thought Prerational? In M. Foster and S. Brandes (eds), *Symbol as Sense.* New York: Academic Press, 25–44.

Sperber, D. (1985a). Anthropology and Psychology: Towards an Epidemiology of Representations (Malinowski Memorial Lecture 1984). *Man*, n.s. 20, 73–89.

Sperber, D. (1985b). *On Anthropological Knowledge.* Cambridge: Cambridge University Press.

Sperber, D. (1986). Issues in the Ontology of Culture. In R. B. Marcus et al. (eds), *Logic, Methodology and Philosophy of Science* , vol. 7. Amsterdam: Elsevier Science Publishers, 557–71.

Sperber, D. (1987). Les Sciences cognitives, les sciences sociales et le matérialisme. *Le Débat*, 47, 105–15.

Sperber, D. (1989). L'Étude anthropologique des représentations: problèmes et perspectives. In D. Jodelet (ed.), *Les Représentations sociales.* Paris: Presses Universitaires de France, 115–30.

Sperber, D. (1990a). The Epidemiology of Beliefs. In C. Fraser and G. Gaskell (eds), *The Social Psychological Study of Widespread Beliefs.* Oxford: Clarendon Press, 25–44.

Sperber, D. (1990b). The Evolution of the Language Faculty: A Paradox and its Solution. *Behavioral and Brain Sciences*, 13(4), 756–8.

Sperber, D. (1991). Culture and Matter. In J.-C. Gardin and C. S. Peebles (eds), *Representations in Archeology.* Bloomington: University of Indiana Press, 56–65.

Sperber, D. (1993). Interpreting and Explaining Cultural Representations. In G. Palsson (ed.), *Beyond Boundaries: Understanding, Translation and Anthropological Discourse*, Oxford: Berg, 162–83.

Sperber, D. (1994). The Modularity of Thought and the Epidemiology of Representations. In L. A. Hirschfeld and S. A. Gelman (eds), *Mapping the Mind: Domain Specificity in Cognition and Culture*. New York: Cambridge University Press, 39–67.

Sperber, D. and Wilson, D. (1986). *Relevance: Communication and Cognition*. Oxford: Blackwell; 2nd edn 1995.

Steiner, F. (1956). *Taboo*. London: Cohen and West.

Stich, S. (1983). *From Folk Psychology to Cognitive Science*. Cambridge, Mass.: MIT Press.

Stich, S. (1990). *The Fragmentation of Reason*. Cambridge, Mass.: MIT Press.

Symons, D. (1979). *The Evolution of Human Sexuality*. New York: Oxford University Press.

Tarde, G. (1895). *Les Lois de l'imitation*. Paris: Félix Alcan.

Tarde, G. (1898). *Les Lois sociales*. Paris: Félix Alcan.

Tooby, J. and Cosmides, L. (1989). Evolutionary Psychology and the Generation of Culture, Part I: Theoretical Considerations. *Ethology & Sociobiology*, 10, 29–49.

Tooby, J. and Cosmides, L. (1992). The Psychological Foundations of Culture. In J. Barkow, L. Cosmides and J. Tooby (eds), *The Adapted Mind: Evolutionary Psychology and the Generation of Culture*. New York: Oxford University Press, 19–136.

Tyler S. A (ed.) (1969). *Cognitive Anthropology*. New York: Holt, Rinehart & Winston.

Vygotsky, L. (1965). *Thought and Language*. Cambridge, Mass.: MIT Press.

Wilensky, R. (1983). Story Grammar versus Story Points. *Behavioral and Brain Sciences*, 6, 579–623.

Williams, G. C. (1966). *Adaptation and Natural Selection*. Princeton: Princeton University Press.

Wilson, E. O. and Bossert, W. (1971). *A Primer of Population Biology*. Sunderland, Mass.: Sinauer.

Wittgenstein, L. (1953). *Philosophical Investigations*. Oxford: Blackwell.

Index